PRAISE FOR
BAD SCIENCE

"Unmissable . . . In a froth of entirely justified indignation, Goldacre slams the mountebanks and bullshitters who misuse science."
—Nigel Hawkes, *The Times* (London), Books of the Year

"Thousands of books are enjoyable; many are enlightening; only a very few will ever rate as necessary to social health. This is one of them." —Boyd Tonkin, *The Independent*

"You'll laugh your head off, then throw all those expensive health foods in the bin."
—*The Observer* (London), Best Books of the Year

"A fine lesson in how to skewer the enemies of reason and the peddlers of cant and half-truths." —*The Economist*

BEN GOLDACRE
BAD SCIENCE

Ben Goldacre is a doctor and science writer who has written the "Bad Science" column in *The Guardian* since 2003. He is thirty-six and lives in London.

BAD SCIENCE

FABER AND FABER, INC.

AN AFFILIATE OF FARRAR, STRAUS AND GIROUX

NEW YORK

BAD SCIENCE

SCIENCE

QUACKS, HACKS,
and BIG PHARMA
. . FLACKS . .

BEN GOLDACRE

Faber and Faber, Inc.
An affiliate of Farrar, Straus and Giroux
18 West 18th Street, New York 10011

Copyright © 2008, 2009, 2010 by Ben Goldacre
All rights reserved
Printed in the United States of America
Originally published, in different form, in 2008 by Fourth Estate, Great Britain
Published in the United States by Faber and Faber, Inc.
First American edition, 2010

Diagrams © HarperCollins*Publishers*, designed by HL Studios, Oxfordshire.

Library of Congress Cataloging-in-Publication Data
Goldacre, Ben.
    Bad science : quacks, hacks, and big pharma flacks / Ben Goldacre.—
1st American ed.
        p.   cm.
    Originally published: London : Fourth Estate, 2008.
    Includes bibliographical references and index.
    ISBN: 978-0-86547-918-0 (pbk. : alk. paper)
    1. Errors, Scientific—Popular works.   2. Pseudoscience—Popular
works.   I. Title.

Q172.5.E77G65 2010
500—dc22

                                                          2010014401

Designed by Jonathan D. Lippincott

www.fsgbooks.com

1   3   5   7   9   10   8   6   4   2

To whom it may concern

CONTENTS

PREFACE

It's easy to laugh at quacks—but this book is not about easy targets or individuals. It follows a natural crescendo, from the foolishness of quacks, via the credence they are given in the mainstream media, through the tricks of the fifty-five-billion-dollar food supplements industry, the evils of the six-hundred-billion-dollar pharmaceuticals industry, the tragedy of science reporting, and on to cases where people have wound up in prison, derided, or dead, simply through the poor understanding of statistics and evidence that pervades our society.

At the time of C. P. Snow's famous lecture on the two cultures of science and the humanities half a century ago, arts graduates simply ignored us. Today, scientists and doctors find themselves outnumbered and outgunned by vast armies of individuals who feel entitled to pass judgment on matters of evidence—an admirable aspiration—without troubling themselves to obtain a basic understanding of the issues.

At school you were taught about chemicals in test tubes, equations to describe motion, and maybe something on photosynthesis—about which more later—but in all likelihood you were taught nothing about death, risk, statistics, and the science of what will kill or cure you. The hole in our culture is gaping: evidence-based medicine, the ultimate applied science, contains some of the cleverest ideas from the past two centuries; it has saved millions of lives, but there has never once been a single exhibit on the subject in London's Science Museum.

This is not for a lack of interest. We are obsessed with health—half of all science stories in the media are medical—and are repeatedly bombarded with sciencey-sounding claims and stories. But as you will see, we get our information from the very people who have repeatedly demonstrated themselves to be incapable of reading, interpreting, and bearing reliable witness to the scientific evidence.

Before we get started, let me map out the territory.

First, we will look at what it means to do an experiment, to see the results with your own eyes, and judge whether they fit with a given theory, or whether an alternative is more compelling. You may find these early steps childish and patronizing—the examples are certainly refreshingly absurd—but they all have been promoted credulously and with great authority in the mainstream media. We will look at the attraction of sciencey-sounding stories about our bodies and the confusion they can cause.

Then we will move on to homeopathy, not because it's important or dangerous—it's not—but because it is the perfect model for teaching evidence-based medicine. Homeopathy pills are, after all, empty little sugar pills that seem to work, and so they embody everything you need to know about "fair tests" of a treatment and how we can be misled into thinking that any intervention is more effective than it really is. You will learn all there is to know about how to do a trial properly and how to spot a bad one. Hiding in the background is the placebo effect, probably the most fascinating and misunderstood aspect of human healing, which goes far beyond a mere sugar pill: it is counterintuitive, it is strange, it is the true story of mind-body healing, and it is far more interesting than any made-up nonsense about therapeutic quantum energy patterns. We will review the evidence on its power, and you will draw your own conclusions.

Then we move on to the bigger fish. Nutritionists are alternative therapists but have somehow managed to brand themselves as men and women of science. Their errors are much more interesting than those of the homeopaths, because they have a grain of real science to them, and that makes them not only more interesting but also more dangerous, because the real threat from cranks is not that their customers might die—there is the odd case, although it seems crass to harp on about them—but that they systematically undermine the public's understanding of the very nature of evidence.

We will see the rhetorical sleights of hand and amateurish errors that have led to your being repeatedly misled about food and nutrition, and how this new industry acts as a distraction from the genuine lifestyle risk factors for ill health, as well as its more subtle but equally alarming impact on the way we see ourselves and our bodies, specifically in the widespread move to medicalize social and political problems, to conceive of them in a reductionist, biomedical framework, and peddle commodifiable solutions, particularly in the form of pills and faddish diets. I will show you evidence that a vanguard of startling wrongness is entering British universities, alongside genuine academic research into nutrition. Then we apply these same tools to proper medicine and see the tricks used by the pharmaceutical industry to pull the wool over the eyes of doctors and patients.

Next we will examine how the media promote the public misunderstanding of science, their single-minded passion for pointless nonstories, and their basic misunderstandings of statistics and evidence, which illustrate the very core of why we do science: to prevent ourselves from being misled by our own atomized experiences and prejudices. Finally, in the part of the book I find most worrying, we will see how people in positions of great power, who should know better, still commit basic errors, with grave consequences, and we will see how the media's cynical distortion of evidence in two specific health scares reached dangerous and frankly grotesque extremes. It's your job to notice, as we go, how incredibly prevalent this stuff is, but also to think what you might do about it.

You cannot reason people out of positions they didn't reason themselves into. But by the end of this book you'll have the tools to win—or at least understand—any argument you choose to initiate, whether it's on miracle cures, MMR, the evils of big pharma, the likelihood of a given vegetable preventing cancer, the dumbing down of science reporting, dubious health scares, the merits of anecdotal evidence, the relationship between body and mind, the science of irrationality, the lexicalization of everyday life, and more. You'll have seen the evidence behind some very popular deceptions, but along the way you'll also have picked up everything useful there is to know about research, levels of evidence, bias, statistics (relax), the history of science, antiscience movements and quackery, and fallen over just some of the amazing stories that the natural sciences can tell us about the world along the way.

It won't be even slightly difficult, because this is the only science lesson where I can guarantee that the people making the stupid mistakes won't be you. And if, by the end, you reckon you might still disagree with me, then I offer you this: you'll still be wrong, but you'll be wrong with a lot more panache and flair than you could possibly manage right now.

BAD SCIENCE

1

MATTER

I spend a lot of time talking to people who disagree with me—I would go so far as to say that it's my favorite leisure activity—and repeatedly I meet individuals who are eager to share their views on science despite the fact that they have *never done an experiment*. They have never tested an idea for themselves, using their own hands, or seen the results of that test, using their own eyes, and they have never thought carefully about what those results mean for the idea they are testing, using their own brain. To these people "science" is a monolith, a mystery, and an authority, rather than a method.

Dismantling our early, more outrageous pseudoscientific claims is an excellent way to learn the basics of science, partly because science is largely about disproving theories, but also because the lack of scientific knowledge among miracle cure artistes, marketers, and journalists gives us some very simple ideas to test. Their knowledge of science is rudimentary, so as well as making basic errors of reasoning, they rely on notions like magnetism, oxygen, water, "energy," and toxins—ideas from high school-level science and all very much within the realm of kitchen chemistry.

DETOX AND THE THEATER OF GOO

Since you'll want your first experiment to be authentically messy,
we'll start with detox. Detox footbaths have been promoted un-
critically in some very embarrassing articles in the New York *Daily
News*, the *Telegraph*, the *Mirror*, *The Sunday Times* (London), GQ
magazine, and various TV shows. Here is a taster from the New
York *Daily News*: it's a story about Ally Shapiro, a fourteen-year-
old who went to a "detox" center run by Roni DeLuz, author of
21 Pounds in 21 Days: The Martha's Vineyard Diet.

"The first day I did it," says Shapiro, "the water was completely black by the end." By day three, twenty minutes in the footbath generated a copper-colored sludge—the color of the flushed
buildup from her joints related to arthritis, DeLuz explained. The
hypothesis from these companies is very clear: your body is full of
"toxins," whatever those may be; your feet are filled with special
"pores" (discovered by ancient Chinese scientists, no less); you
put your feet in the bath, the toxins are extracted, and the water
goes brown. Is the brown in the water because of the toxins? Or is
that merely theater?

One way to test this is to go along and have an Aqua Detox
treatment yourself at a health spa, beauty salon, or any of the
thousands of places they are available online, and take your feet
out of the bath when the therapist leaves the room. If the water
goes brown without your feet in it, then it wasn't your feet or your
toxins that did it. That is a controlled experiment; everything is
the same in both conditions, except for the presence or absence of
your feet.

There are disadvantages with this experimental method (and
there is an important lesson here—that we must often weigh up the
benefits and practicalities of different forms of research, which will
become important in later chapters). From a practical perspective,

the "feet out" experiment involves subterfuge, which may make you uncomfortable. But it is also expensive: one session of Aqua Detox will cost more than the components to build your own detox device, a perfect model of the real one.

You will need:

- One car battery charger
- Two large nails
- Kitchen salt
- Warm water
- One Barbie doll
- A full analytic laboratory (optional)

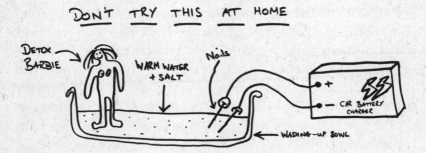

This experiment involves electricity and water. In a world of hurricane hunters and volcanologists, we must accept that everyone sets their own level of risk tolerance. You might well give yourself a nasty electric shock if you perform this experiment at home, and it could easily blow the wiring in your house. It is not safe, but it is in some sense relevant to your understanding of MMR, homeopathy, postmodernist critiques of science, and the evils of big pharma. DO NOT BUILD IT.

When you switch your Barbie Detox machine on, you will see that the water goes brown, due to a very simple process called electrolysis; the iron electrodes rust, essentially, and the brown rust goes into the water. But there is something more happening in there, something you might half remember from chemistry at school. There is salt in the water. The proper scientific term for household salt is "sodium chloride"; in solution, this means that there are chloride ions floating around, which have a negative charge (and sodium ions, which have a positive charge). The red connector on your car battery charger is a "positive electrode," and here negatively charged electrons are stolen away from the negatively charged chloride ions, resulting in the production of free chlorine gas.

So chlorine gas is given off by the Barbie Detox bath, and indeed by the Aqua Detox footbath, and the people who use this product have elegantly woven that distinctive chlorine aroma into their story: it's the chemicals, they explain; it's the chlorine coming out of your body, from all the plastic packaging on your food and all those years bathing in chemical swimming pools. "It has been interesting to see the color of the water change and smell the chlorine leaving my body," says one testimonial for the similar product Emerald Detox. At another sales site: "The first time she tried the Q2 [Energy Spa], her business partner said his eyes were burning from all the chlorine that was coming out of her, leftover [sic] from her childhood and early adulthood." All that chemically chlorine gas that has accumulated in your body over the years. It's a frightening thought.

But there is something else we need to check. Are there toxins in the water? Here we encounter a new problem: What do they mean by toxins? I've asked the manufacturers of many detox products this question time and again, but they demur. They wave their hands, they talk about stressful modern lifestyles, they talk about pollution, they talk about junk food, but they will not tell me the name of a single chemical that I can measure. "What toxins are being extracted from the body with your treatment?" I ask. "Tell me what is in the water, and I will look for it in a laboratory." I have never been given an answer.

After much of their hedging and fudging, I chose two chemicals pretty much at random: creatinine and urea. These are common breakdown products from your body's metabolism, and your kidneys get rid of them in urine. Through a friend, I went for a genuine Aqua Detox treatment, took a sample of brown water, and used the disproportionately state-of-the-art analytic facilities of St. Mary's Hospital in London to hunt for these two chemical "toxins." There were no toxins in the water. Just lots of brown, rusty iron.

Now, with findings like these, scientists might take a step back and revise their ideas about what is going on with the footbaths. We don't really expect the manufacturers to do that, but what they say in response to these findings is very interesting, at least to me, because it sets up a pattern that we will see repeated throughout the world of pseudoscience: instead of addressing the criticisms, or embracing the new findings in a new model, they seem to shift the goalposts and retreat, crucially, into *untestable positions*.

Some of them now deny that toxins come out in the footbath (which would stop me measuring them); your body is somehow informed that it is time to release toxins in the normal way—whatever that is, and whatever the toxins are—only more so. Some of them now admit that the water goes a bit brown without your feet in it, but "not as much." Many of them tell lengthy stories about the "bioenergetic field," which they say cannot be measured except by how well you are feeling. All of them talk about how stressful modern life is.

That may well be true. But it has nothing to do with their footbath, which is all about theater, and theater is the common theme for all detox products, as we shall see. On with the brown goo.

EAR CANDLES

You might think that Hopi ear candles are easy targets. But their efficacy has still been cheerfully promoted by *The Independent*,

The Observer, and the BBC, to name just a few respected British news outlets. They pop up endlessly in American local papers desperate to fill space, from the *Alameda Times-Star* to the *Syracuse Post-Standard*. Since journalists like to present themselves as authoritative purveyors of scientific information, I'll let the internationally respected BBC explain how these hollow wax tubes, Hopi ear candles, will detox your body: "The candles work by vaporizing their ingredients once lit, causing convectional air flow towards the first chamber of the ear. The candle creates a mild suction which lets the vapors gently massage the eardrum and auditory canal. Once the candle is placed in the ear it forms a seal which enables wax and other impurities to be drawn out of the ear." The proof comes when you open a candle up and discover that it is filled with a familiar waxy orange substance, which must surely be earwax. If you'd like to test this yourself, you will need: an ear, a clothespin, some poster putty, a dusty floor, some scissors, and two ear candles.

If you light one ear candle, and hold it over some dust, you will find little evidence of any suction. Before you rush to publish your finding in a peer-reviewed academic journal, someone has beaten you to it: a paper published in the medical journal *Laryngoscope* used expensive tympanometry equipment and found— as you have—that ear candles exert no suction. There is no truth to the claim that doctors dismiss alternative therapies out of hand.

But what if the wax and toxins are being drawn into the candle by some other, more esoteric route, as is often claimed?

For this you will need to do something called a controlled experiment, comparing the results of two different situations, where one is the experimental condition, the other is the control condition, and the only difference is the thing you're interested in testing. This is why you have two candles.

Put one ear candle in someone's ear, as per the manufacturer's instructions, and leave it there until it burns down.* Put the other candle in the clothespin, and stand it upright using the Blu-Tack; this is the "control arm" in your experiment. The point of a control is simple: we need to minimize the differences between the two setups, so that the only real difference between them is the single factor you're studying, which in this case must be: "Is it my ear that produces the orange goo?"

Take your two candles back inside and cut them open. In the "ear" candle, you will find a waxy orange substance. In the "picnic table control," you will find a waxy orange substance. There is only one internationally recognized method for identifying something as earwax: pick some up on the end of your finger, and touch it with your tongue. If your experiment had the same results as mine, both of them taste a lot like candle wax.

Does the ear candle remove earwax from your ears? You can't tell, but a published study followed patients during a full program of ear candling and found no reduction. For all that you might have learned something useful here about the experimental method, there is something more significant you should have picked up: it is expensive, tedious, and time-consuming to test every whim concocted out of thin air by therapists selling unlikely miracle cures. But it can be done, and it is done.

DETOX PATCHES AND THE HASSLE BARRIER

Last in our brown sludge detox triptych comes the detox foot patch. These are available in most health food stores or from your local Avon lady (this is true). They look like teabags, with a foil backing that you stick onto your foot using an adhesive edging, before you get into bed. When you wake up the next morning, there is a

---

*Be careful. One paper surveyed 122 ENT doctors and collected twenty-one cases of serious injury from burning wax falling onto the eardrum during ear candle treatment.

strange-smelling, sticky brown sludge attached to the bottom of your foot and inside the teabag. This sludge—you may spot a pattern here—is said to be "toxins." Except it's not. By now you can probably come up with a quick experiment to show that. I'll give you one option in a footnote.*

An experiment is one way of determining whether an observable effect—sludge—is related to a given process. But you can also pull things apart on a more theoretical level. If you examine the list of ingredients in these patches, you will see that they have been very carefully designed.

The first thing on the list is "pyroligneous acid," or wood vinegar. This is a brown powder that is highly hygroscopic, a word that simply means it attracts and absorbs water, like those little silica bags that come in electronic equipment packaging. If there is any moisture around, wood vinegar will absorb it and make a brown mush that feels warm against your skin.

What is the other major ingredient, impressively listed as "hydrolyzed carbohydrate"? A carbohydrate is a long string of sugar molecules all stuck together. Starch is a carbohydrate, for example, and in your body this is broken down gradually into the individual sugar molecules by your digestive enzymes, so that you can absorb it. The process of breaking down a carbohydrate molecule into its individual sugars is called hydrolysis. So "hydrolyzed carbohydrate," as you might have worked out by now, for all that it sounds sciencey, basically means "sugar." Obviously sugar goes sticky in sweat.

Is there anything more to these patches than that? Yes. There is a new device, which we should call the hassle barrier, another recurring theme in the more advanced forms of foolishness that we shall be reviewing later. There are huge numbers of different

---

*If you take one of these bags and squirt some water onto it, then pop a nice hot cup of tea on top of it and wait for ten minutes, you'll see brown sludge forming. There are no toxins in porcelain.

brands, and many of them offer excellent and lengthy documents full of science to prove that they work: they have diagrams and graphs and the appearance of scienciness, but the key elements are missing. There are experiments, they say, which prove that detox patches do something . . . but they don't tell you what these experiments consisted of, or what their "methods" were; they offer only decorous graphs of "results."

To focus on the methods is to miss the point of these apparent "experiments": they aren't about the methods; they're about the positive result, the graph, and the appearance of science. These are superficially plausible totems to frighten off a questioning journalist, a *hassle barrier*, and this is another recurring theme, which we will see—in more complex forms—around many of the more advanced areas of bad science. You will come to love the details.

IF IT'S NOT SCIENCE, WHAT IS IT?

But there is something important happening here, with detox, and I don't think it's enough just to say, "All this is nonsense." The detox phenomenon is interesting because it represents one of the most grandiose innovations of marketers, lifestyle gurus, and alternative therapists: the invention of a whole new physiological process. In terms of basic human biochemistry, detox is a meaningless concept. It doesn't cleave nature at the joints. There is nothing on the "detox system" in a medical textbook. That burgers and beer can have negative effects on your body is certainly true, for a number of reasons; but the notion that they leave a specific residue, which can be extruded by a specific process, a physiological system called detox, is a marketing invention.

If you look at a metabolic flowchart, the gigantic wall-size maps of all the molecules in your body, detailing the way that food is broken down into its constituent parts, and then those compo-

nents are converted between each other, and then those new building blocks are assembled into muscle, and bone, and tongue, and
bile, and sweat, and booger, and hair, and skin, and sperm, and
brain, and everything that makes you you, it's hard to pick out one
thing that is the "detox system."

Because it has no scientific meaning, detox is much better understood as a cultural product. Like the best pseudoscientific in-
ventions, it deliberately blends useful common sense with
outlandish, medicalized fantasy. In some respects, how much you
buy into this reflects how self-dramatizing you want to be or, in
less damning terms, how much you enjoy ritual in your daily life.
When I go through busy periods of partying, drinking, sleep de-
privation, and convenience eating, I usually decide—eventually—
that I need a bit of a rest. So I have a few nights in, reading at
home, and eating more salad than usual. Models and celebrities,
meanwhile, "detox" with Master Cleanse and the Fruit Flush
Diet.

On one thing we must be absolutely clear, because this is a
recurring theme throughout the world of bad science: there is
nothing wrong with the notion of eating healthily and abstaining
from various risk factors for ill health like excessive alcohol use.
But that is not what detox is about; these are quick-fix health
drives, constructed from the outset as short term, while lifestyle
risk factors for ill health have their impact over a lifetime. But I
am even willing to agree that some people might try a five-day
detox and remember (or even learn) what it's like to eat vegetables,
and that gets no criticism from me.

What's wrong is to pretend that these rituals are based in science or even that they are new. Almost every religion and culture
have some form of purification or abstinence ritual, with fasting, a
change in diet, bathing, or any number of other interventions,
most of which are dressed up in mumbo jumbo. They're not presented as science, because they come from an era before scientific

terms entered the lexicon, but still: Yom Kippur in Judaism, Ramadan in Islam, and all manner of other similar rituals in Christianity, Hinduism, the Baha'i faith, Buddhism, and Jainism are each about abstinence and purification (among other things). Such rituals, like detox regimes, are conspicuously and—to some believers too, I'm sure—spuriously precise. Hindu fasts, for example, if strictly observed, run from the previous day's sunset until *forty-eight minutes* after the next day's sunrise.

Purification and redemption are such recurrent themes in ritual because there is a clear and ubiquitous need for them; we all do regrettable things as a result of our own circumstances, and new rituals are frequently invented in response to new circumstances. In Angola and Mozambique, purification and cleansing rituals have arisen for children affected by war, particularly former child soldiers. These are healing rituals, in which the child is purged and purified of sin and guilt, of the "contamination" of war and death (contamination is a recurring metaphor in all cultures, for obvious reasons); the child is also protected from the consequences of his previous actions, which is to say, he is protected from retaliation by the avenging spirits of those he has killed. As a World Bank report put it in 1999:

> These cleansing and purification rituals for child soldiers
> have the appearance of what anthropologists call rites of
> transition. That is, the child undergoes a symbolic change
> of status from someone who has existed in a realm of sanc-
> tioned norm-violation or norm-suspension (i.e. killing,
> war) to someone who must now live in a realm of peaceful
> behavioral and social norms, and conform to these.

I don't think I'm stretching this too far. In what we call the developed Western world, we seek redemption and purification from the more extreme forms of our material indulgence: we fill our faces with drugs, drink, bad food, and other indulgences, we know it's wrong, and we crave ritualistic protection from the consequences, a public "transitional ritual" commemorating our return to healthier behavioral norms.

The presentation of these purification diets and rituals has always been a product of their time and place, and now that science is our dominant explanatory framework for the natural and moral world, for right or wrong, it's natural that we should bolt a bastardized pseudoscientific justification onto our redemption. Like so much of the nonsense in bad science, "detox" pseudoscience isn't something done *to* us, by venal and exploitative outsiders; it is a cultural product, a recurring theme, and we do it to ourselves.

2

BRAIN GYM

Under normal circumstances this should be the part of the book
where I fall into a rage over creationism, to gales of left-wing ap-
plause. But if you want an example that's less covered in the media,
there is a vast empire of pseudoscience being peddled, for hard cash,
in such liberal enclaves in the United States as Boulder, Colorado,
and Portland, Oregon. It's even more successful abroad, where it has
been taught in more than eighty countries and made it into public
schools up and down the United Kingdom.* It's called Brain Gym,
it's an export from the United States (thanks), it's swallowed whole
by teachers, it's presented directly to the children they teach, and it's
riddled with transparent, shameful, and embarrassing nonsense.

At the heart of Brain Gym is a string of complicated and proprietary exercises for kids that "enhance the experience of whole

---

*Though it's an American company, Brain Gym hasn't quite caught on in the United
States to the same degree that it has in the U.K. However, at least two American uni-
versities offer accreditation. You can get graduate credit for salary upgrade and recerti-
fication from the University of Colorado at Denver for Brain Gym and ADHD classes.
For classes taught outside Colorado or adjoining states, you can get CEUs from Dominican
University of California. Also Grand Valley State University.

brain learning." They're very keen on water, for example. "Drink a glass of water before Brain Gym activities," they say. "As it is a major component of blood, water is vital for transporting oxygen to the brain." Heaven forbid that your blood should dry out. This water should be held in your mouth, they say, because then it can be absorbed *directly* from there into your brain.

Is there anything else you can do to get blood and oxygen to your brain more efficiently? Yes, an exercise called "Brain Buttons": "Make a 'C' shape with your thumb and forefinger and place on either side of the breastbone just below the collarbone. Gently rub for twenty or thirty seconds whilst placing your other hand over your navel. Change hands and repeat. This exercise stimulates the flow of oxygen carrying blood through the carotid arteries to the brain to awaken it and increase concentration and relaxation." Why? "Brain buttons lie directly over and stimulate the carotid arteries."

Children can be disgusting, and often they can develop extraordinary talents, but I've yet to meet any child who can stimulate his carotid arteries inside his rib cage. That's probably going to need the sharp scissors that only Mommy can use.

You might imagine that this nonsense is a marginal, peripheral trend that I have contrived to find in a small number of isolated, misguided schools. But no. Brain Gym is practiced in hundreds, if not thousands, of mainstream state schools throughout the U.K. As of today I have a list of over four hundred schools that mention it specifically by name on their websites, and many, many others will also be using it. Ask if they do it at your school. I'd be genuinely interested to know their reaction.

Perhaps if they could just do the "hook-up" exercises on page 31 of the *Brain Gym Teacher's Manual* (where you press your fingers against each other in odd contorted patterns), this would "connect the electrical circuits in the body, containing and thus focusing both attention and disorganized energy," and they would finally see sense. Perhaps if they wiggled their ears with their fingers as per the Brain Gym textbook, it would "stimulate the reticular formation of the brain to tune out distracting, irrelevant sounds and tune into language."

The same teacher who explains to children how blood is pumped around the lungs and then the body by the heart is also telling them that when they do the "Energizer" exercise (which is far too complicated to describe), "this back and forward movement of the head increases the circulation to the frontal lobe for greater comprehension and rational thinking." Most frighteningly, this teacher sat through a class being taught this nonsense by a Brain Gym instructor, without challenging or questioning it.

In some respects the issues here are similar to those in the chapter on detox: if you just want to do a breathing exercise, then that's great. But the creators of Brain Gym go much further. Their special, proprietary, theatrical yawn will lead to "increased oxidation for efficient relaxed functioning." Oxidation is what causes rusting. It is not the same as oxygenation, which I suppose is what they mean. (And even if they are talking about oxygenation, you don't need to do a funny yawn to get oxygen into your blood: like most other wild animals, children have a perfectly adequate and fascinating physiological system in place to regulate their blood oxygen and carbon dioxide levels, and I'm sure many of them would rather be taught about that, and indeed about the role of electricity in the body, or any of the other things Brain Gym confusedly jumbles up, than this transparent pseudoscientific nonsense.)

How can this nonsense be taught in schools? One obvious explanation is that the teachers have been blinded by all these clever long phrases like "reticular formation" and "increased oxidation." As it happens, this very phenomenon has been studied in a fascinating set of experiments from the March 2008 edition of the *Journal of Cognitive Neuroscience*, which elegantly demonstrated that people will buy into bogus explanations much more readily when they are dressed up with a few technical words from the world of neuroscience.

Subjects were given descriptions of various phenomena from the world of psychology and then randomly offered one of four explanations for them. The explanations either contained neuroscience or didn't, and were either "good" explanations or "bad" ones (bad ones being, for example, simply circular restatements of the phenomenon itself or empty words).

Here is one of the scenarios. Experiments have shown that people are quite bad at estimating the knowledge of others; if *we* know the answer to a question about a piece of trivia, we overestimate the extent to which other people will know that answer too. In the experiment a "without neuroscience" explanation for this phenomenon was: "The researchers claim that this [overestimation] happens because subjects have trouble switching their point of view to consider what someone else might know, mistakenly projecting their own knowledge onto others." (This was a "good" explanation.)

A "with neuroscience" explanation—and a cruddy one too—was this: "Brain scans indicate that this [overestimation] happens because of the frontal lobe brain circuitry known to be involved in self-knowledge. Subjects make more mistakes when they have to judge the knowledge of others. People are much better at judging what they themselves know." Very little is added by this explanation, as you can see. Furthermore, the neuroscience information is merely decorative and irrelevant to the explanation's logic.

The subjects in the experiment were from three groups—everyday people, neuroscience students, and neuroscience academics—and they performed very differently. All three groups judged good explanations as more satisfying than bad ones, but the subjects in the two nonexpert groups judged that the explanations *with* the logically irrelevant neurosciencey information were more satisfying than the explanations *without* the spurious neuroscience. What's more, the spurious neuroscience had a particularly strong effect on people's judgments of "bad" explanations. Quacks, of course, are well aware of this and have been adding sciencey-sounding explanations to their products for as long as quackery has existed, as a means to bolster their authority over the patient (in an era, interestingly, when doctors have struggled to inform patients more and to engage them in decisions about their own treatment).

It's interesting to think about why this kind of decoration is so seductive, and to people who should know better. First, the very presence of neuroscience information might be seen as a surrogate marker of a "good" explanation, regardless of what is actually said. As the researchers say, "Something about seeing neuroscience information may encourage people to believe they have received a scientific explanation when they have not."

But more clues can be found in the extensive literature on irrationality. People tend, for example, to rate longer explanations as being more similar to "experts' explanations." There is also the "seductive details" effect: if you present related (but logically irrelevant) details to people as part of an argument, this seems to make it more difficult for them to encode, and later recall, the main argument of a text, because their attention is diverted.

More than this, perhaps we all fall for reductionist explanations about the world. They just feel neat somehow. When we read the neurosciencey language in the "bogus neuroscience explanations" experiment—and in the Brain Gym literature—we feel as if we have been given a physical explanation for a behavioral phenomenon ("an exercise break in class is refreshing"). We have somehow made behavioral phenomena feel connected to a larger explanatory system, the physical sciences, a world of certainty, graphs, and unambiguous data. It feels like progress. In fact, as is often the case with spurious certainty, it's the very opposite.

Again, we should focus for a moment on what is good about Brain Gym, because when you strip away the nonsense, it advocates regular breaks, intermittent light exercise, and drinking plenty of water. This is all entirely sensible.

But Brain Gym perfectly illustrates two more recurring themes from the industry of pseudoscience. The first is this: you can use hocus pocus—or what Plato euphemistically called a noble myth— to make people do something fairly sensible like drink some water and have an exercise break. You will have your own view on when this is justified and proportionate (perhaps factoring in issues like whether it's necessary and the side effects of pandering to non-sense), but it strikes me that in the case of Brain Gym, this is not a close call: children are predisposed to learn about the world from adults, and specifically from teachers; they are sponges for informa-tion, for ways of seeing, and authority figures who fill their heads with nonsense are sowing the ground, I would say, for a lifetime of exploitation.

The second theme is perhaps more interesting: the proprieto-rialization of common sense. You can take a perfectly sensible in-tervention, like a glass of water and an exercise break, but add nonsense, make it sound more technical, and make yourself sound clever. This will enhance the placebo effect, but you might also wonder whether the primary goal is something much more cyni-cal and lucrative: to make common sense copyrightable, unique, patented, and *owned*.

We will see this time and again, on a grander scale, in the work of dubious health care practitioners and specifically in the field of "nutritionism," because scientific knowledge—and sensible dietary advice—are free and in the public domain. Anyone can use it, un-derstand it, sell it, or simply give it away. Most people know what constitutes a healthy diet already. If you want to make money out of it, you have to make a space for yourself in the market, and to do this, you must overcomplicate it, attach your own dubious stamp.

Is there any harm in this process? Well, it's certainly wasteful, and it does seem peculiar to give money away for basic diet advice or exercise breaks at school. But there are other hidden dangers, which are far more corrosive. This process of professionalizing the obvious fosters a sense of mystery around science and health advice that is unnecessary and destructive. More than anything, more than the unnecessary ownership of the obvious, it is disempowering. All too often this spurious privatization of common sense is happening in areas where we could be taking control, doing it ourselves, feeling our own potency and our ability to make sensible decisions; instead we are fostering our dependence on expensive outside systems and people.

But what's most frightening is the way that pseudoscience makes your head go soggy. Debunking Brain Gym, let me remind you, does not require high-end, specialist knowledge. We are talking about a program that claims that "processed foods do not contain water," possibly the single most rapidly falsifiable statement I've seen all week. What about soup? "All other liquids are processed in the body as food, and do not serve the body's water needs."

This is an organization at the edges of reason, but it is operating in countless schools. When I wrote about Brain Gym in my U.K. newspaper column in 2005, saying that "exercise breaks good, pseudoscientific nonsense laughable," while many teachers erupted with delight, many were outraged and "disgusted" by what they decided was an attack on exercises that they experienced as helpful. One— an assistant head teacher no less—demanded: "From what I can gather you have visited no classrooms, interviewed no teachers nor questioned any children, let alone had a conversation with any of a number of specialists in this field?" Do I need to visit a classroom to find out if there is water in processed food? No. If I meet a "specialist" who tells me that a child can massage both carotid arteries through the rib cage (without scissors), what will I say to him? If I meet a teacher who thinks that touching your fingers together will connect the electrical circuit of the body, where do we go from there?

I'd like to imagine that teachers might have the common sense to spot this nonsense and stop it in its tracks. Just one thing gives me hope, and that is the steady trickle of e-mails I receive on the subject from children, ecstatic with delight at the stupidity of their teachers:

> I'd like to submit to Bad Science my teacher who gave us a handout which says that "Water is best absorbed by the body when provided in frequent small amounts." What I want to know is this. If I drink too much in one go, will it leak out of my arsehole instead?
>
> "Anton," 2006

Thank you, Anton.

3

THE PROGENIUM
XY COMPLEX

I have great respect for the manufacturers of cosmetics. They are at the other end of the spectrum from the detox industry: this is a tightly regulated industry, with big money to be made from nonsense, and so we find large, well-organized teams from international biotech firms generating elegant, distracting, suggestive, but utterly defensible pseudoscience. After the childishness of Brain Gym, we can now raise our game.

Before we start, it's important to understand how cosmetics—specifically moisturizing creams—actually work, because there should be no mystery here. First, you want your expensive cream to hydrate your skin. They all do that, and Vaseline does the job very well; in fact, much of the important early cosmetics research was about preserving the moisturizing properties of Vaseline, while avoiding its greasiness, and this technical mountain was scaled several decades ago. A thirteen-ounce tub at about five dollars from your local drugstore will do the job excellently.

If you really want to, you can replicate this by making your own moisturizer at home; you're aiming for a mix of water and oil, but one that's "emulsified," which is to say, nicely mixed up. When

I was involved in hippie street theater—and I'm being entirely serious here—we made moisturizer from equal parts of olive oil, coconut oil, honey, and rose water (tap water is fine too). Beeswax is better than honey as an emulsifier, and you can modify the cream's consistency for yourself: more beeswax will make it firmer, more oil will make it softer, and more water makes it sort of fluffier but increases the risk of the ingredients separating out. Get all your ingredients lightly heated, but separately, stir the oil into the wax, beating all the time, and then stir in the water. Stick it in a jar, and keep for three months in the fridge.

The creams in your local pharmacy seem to go way beyond this. They are filled with magic ingredients: Regenium XY technology, Nutrileum complex, RoC Retinol Correxion, VitaNiacin, Covabeads, ATP Stimuline, and Tenseur Peptidique Végétal. Surely you could never replicate that in your kitchen, or with creams that cost as much by the gallon as these ones cost for a squirt of the tiny tube? What are these magic ingredients? And what do they do?

There are basically three groups of ingredients in moisturizing cream. First, there are powerful chemicals, like alpha hydroxy acids, high levels of vitamin C, or molecular variations on the theme of vitamin A. These have genuinely been shown to make your skin seem more youthful, but they are only effective at such high concentrations, or high acidity levels, that the creams cause irritation, stinging, burning, and redness. They were the great white hope in the 1990s, but now they've all had to be massively watered down by law, unless on prescription. No free lunch, and no effects without side effects, as usual.

Companies still name them on the label, wallowing in the glory of their efficacy at higher potencies, because you don't have to give the doses of your ingredients, only their ranked order. But these chemicals are usually in your cream at talismanic concentrations, for show only. The claims made on the various bottles and tubes are from the halcyon days of effective and high-potency acidic creams, but that's hard to tell, because they are usually based on privately funded and published studies, done by the industry, and rarely available in their complete published forms, as a proper academic paper should be, so that you can check the working. Of course, you have to forget that technical stuff; most of the "evidence" quoted in cream ads is from subjective reports, in which "seven out of ten people who received free pots of cream were very pleased with the results." You don't need anybody's help spotting how useless that is as evidence.

The second ingredient in almost all fancy creams is one that does kind of work: cooked and mashed-up vegetable protein (hydrolyzed X-microprotein nutricomplexes, Tenseur Peptidique Végétal, or whatever they're calling them this month). These are long, soggy chains of amino acids, which swim around in the cream, languorously stretched out in the moistness of it all. When the cream dries on your face, these long, soggy chains contract and tighten; the slightly unpleasant taut sensation you get on your face when you wear these creams is from the protein chains contracting all over your skin, which temporarily shrinks your finer wrinkles. It is a fleeting but immediate payoff from using the expensive creams, but it wouldn't help you choose between them, since almost all of them contain mashed-up protein chains.

Finally, there is the huge list of esoteric ingredients, tossed in on a prayer, with suggestive language elegantly woven around them in a way that allows you to believe that all kinds of claims are being made.

Classically, cosmetics companies will take highly theoretical, textbookish information about the way that cells work—the components at a molecular level or the behavior of cells in a glass dish—and then pretend it's the same as the ultimate issue of whether something makes you look nice. "This molecular component," they say, with a flourish, "is crucial for collagen formation."

And that will be perfectly true (along with many other amino acids which are used by your body to assemble protein in joints, skin, and everywhere else), but there is no reason to believe that anyone is deficient in it or that smearing it on your face will make any difference to your appearance. In general, you don't absorb things very well through your skin, because its purpose is to be relatively impermeable. When you sit in a bath of baked beans for charity, you do not get fat, nor do you start farting.

Despite this, on any trip to the pharmacy or department store beauty counter (I recommend it) you can find a phenomenal array of magic ingredients on the market. Valmont Cellular DNA Complex is made from "specially treated salmon roe DNA," but it's spectacularly unlikely that DNA—a very large molecule indeed—would be absorbed by your skin, or indeed be any use for the synthetic activity happening in it, even if it were. You're probably not short of the building blocks of DNA in your body. There's a hell of a lot of it in there already.

Thinking through: if salmon DNA *were* absorbed whole by your skin, then you would be absorbing alien, or rather fish, design blueprints into your cells—that is, the instructions for making fish cells, which might not be helpful for you as a human. It would also be a surprise if the DNA were digested into its constituent elements in your skin (your gut, though, is specifically adapted for digesting large molecules, using digestive enzymes that break them up into their constituent parts before absorption).

The simple theme running through all these products is that you can hoodwink your body, when in reality there are finely tuned "homeostatic" mechanisms, huge, elaborate systems with feedback and measuring devices, constantly calibrating and recalibrating the amounts of various different chemical constituents being sent to different parts of your body. If anything, interfering with that system is likely to have the opposite of the simplistic effects claimed.

As the perfect example, there are huge numbers of creams (and other beauty treatments) claiming to deliver oxygen directly to your skin. Many of the creams contain peroxide, which, if you really want to persuade yourself of its efficacy, has a chemical formula of H_2O_2 and could fancifully be conceived of as water "with some extra oxygen," although chemical formulas don't really work that way; after all, a pile of rust is an iron bridge "with some extra oxygen," and you wouldn't imagine it would oxygenate your skin.

Even if we give them the benefit of the doubt and pretend that these treatments really will deliver oxygen to the surface of the skin, and that this will penetrate meaningfully into the cells, what good would that do? Your body is constantly monitoring the amount of blood and nutrients it's supplying to tissues and the quantity of tiny capillary arteries feeding a given area, and more vessels will grow toward areas with low oxygen, because that is a good index of whether more blood supply is needed. Even if the claim about oxygen in cream's penetrating your tissues were true, your body would simply downregulate the supply of blood to that part of skin, scoring a homeostatic own goal. In reality, hydrogen peroxide is simply a corrosive chemical that gives you a light chemical burn at low strengths. This might explain that fresh, glowing feeling.

These details generalize to most of the claims made on packaging. Look closely at the label or advertisement, and you will routinely find that you are being played in an elaborate semantic game, with the complicity of the regulators. It's rare to find an explicit claim: that rubbing this particular magic ingredient on your face will make you look better. The claim is made for the cream *as a whole*, and it is true for the cream as a whole, because as you now know, all moisturizing creams—even the cheap kinds—will moisturize.

Once you know this, shopping becomes marginally more interesting. The link between the magic ingredient and efficacy is made only in the customer's mind, and reading through the manufacturer's claims, you can see that they have been carefully reviewed by a small army of consultants to ensure that the label is highly suggestive, but also—to the eye of an informed pedant—semantically and legally watertight. (If you want to make a living in this field, I would recommend the well-trodden career path—a spell in trading standards, advertising standards, or any other regulatory body—before going on to work as a consultant to industry.)

So what's wrong with this kind of spin? We should be clear on one thing: I'm not on a consumer crusade. Just like the lottery, the cosmetics industry is playing on people's dreams, and people are free to waste their money. I can very happily view fancy cosmetics—and other forms of quackery—as a special, self-administered, voluntary tax on people who don't understand science properly. I would also be the first to agree that people don't buy expensive cosmetics simply because they have a belief in their efficacy, because it's "a bit more complicated than that": these are luxury goods, status items, and they are bought for all kinds of interesting reasons.

But it's not entirely morally neutral. First, the manufacturers of these products sell shortcuts to smokers and the obese; they sell the idea that a healthy body can be attained by using expensive potions, rather than by simple old-fashioned exercise and eating your greens. This is a recurring theme throughout the world of bad science.

More than that, these ads sell a dubious worldview. They sell the idea that science is not about the delicate relationship between evidence and theory. They suggest, instead, with all the might of their international advertising budgets, their Microcellular Complexes, their Neutrilium XY, their Tenseur Peptidique Végétal, and the rest, that science is about impenetrable nonsense involving equations, molecules, sciencey diagrams, sweeping didactic statements from authority figures in white coats, and that this sciencey-sounding stuff might just as well be made up, concocted, confabulated out of thin air, in order to make money. They sell the idea that science is incomprehensible, with all their might, and they sell this idea mainly to attractive young women, who are disappointingly underrepresented in the sciences.

In fact, they sell the worldview of Teen Talk Barbie from Mattel, who shipped with a sweet little voice circuit inside her so she could say things like "Math class is tough!," "I love shopping!," and "Will we ever have enough clothes?" when you pressed her buttons. In December 1992 the feminist direct-action Barbie Liberation Organization switched the voice circuits of hundreds of Teen Talk Barbies and G.I. Joe dolls in American stores. On Christmas Day Barbie said, "Dead men tell no lies," in a nice assertive voice, and the boys got soldiers under the tree telling them, "Math class is tough!" and asking, "Wanna go shopping?"

The work of the BLO is not yet done.

4

HOMEOPATHY

And now for the meat. But before we take a single step into this arena, we should be clear on one thing: despite what you might think, I'm not desperately interested in complementary and alternative medicine (a dubious piece of phraseological rebranding in itself). I am interested in the role of medicine, our beliefs about the body and healing, and I am fascinated—in my day job—by the intricacies of how we can gather evidence for the benefits and risks of a given intervention.

Homeopathy, in all of this, is simply our tool.

So here we address one of the most important issues in science: How do we know if an intervention works? Whether it's a face cream, a detox regime, a school exercise, a vitamin pill, a parenting program, or a heart attack drug, the skills involved in testing an intervention are all the same. Homeopathy makes the clearest teaching device for evidence based medicine for one simple reason: homeopaths give out little sugar pills, and pills are the easiest thing in the world to study.

By the end of this section you will know more about evidence-based medicine and trial design than the average doctor.

You will understand how trials can go wrong and give false positive results, how the placebo effect works, and why we tend to overestimate the efficacy of pills. More important, you will also see how a health myth can be created, fostered, and maintained by the alternative medicine industry, using all the same tricks on you, the public, that big pharma uses on doctors. This is about something much bigger than homeopathy.

WHAT IS HOMEOPATHY?

Homeopathy is perhaps the paradigmatic example of an alternative therapy. It claims the authority of a rich historical heritage, but its history is routinely rewritten for the PR needs of a contemporary market; it has an elaborate and sciencey-sounding framework for how it works, without scientific evidence to demonstrate its veracity; and its proponents are quite clear that the pills will make you better, when in fact they have been thoroughly researched, with innumerable trials, and have been found to perform no better than placebo.

Homeopathy was devised by a German doctor named Samuel Hahnemann in the late eighteenth century. At a time when mainstream medicine consisted of bloodletting, purging, and various other ineffective and dangerous evils, when new treatments were conjured up out of thin air by arbitrary authority figures who called themselves doctors, often with little evidence to support them, homeopathy would have seemed fairly reasonable.

Hahnemann's theories differed from the competition because he decided—and there's no better word for it—that if he could find a substance that would induce the symptoms of a disease in a healthy individual, it could be used to treat the same symptoms in a sick person. His first homeopathic remedy was cinchona bark, which was suggested as a treatment for malaria. He took some himself, at a high dose, and experienced symptoms that he decided were similar to those of malaria itself: "My feet and finger-tips at once became cold; I grew languid and drowsy; my heart began to palpitate; my pulse became hard and quick; an intolerable anxiety and trembling arose . . . prostration . . . pulsation in the head, redness in the cheek and raging thirst . . . intermittent fever . . . stupefaction . . . rigidity . . ." and so on.

Hahnemann assumed that everyone would experience these symptoms if they took cinchona (although there's some evidence that he just experienced an idiosyncratic adverse reaction). More important, he also decided that if he gave a tiny amount of cinchona to someone with malaria, it would treat, rather than cause, the malaria symptoms. The theory of like cures like, which he conjured up on that day, is, in essence, the first principle of homeopathy.*

Giving out chemicals and herbs could be a dangerous business, since they can have genuine effects on the body (they induce symptoms, as Hahnemann identified). But he solved that problem with his second great inspiration, and the key feature of homeopathy that most people would recognize today: he decided—again, that's the only word for it—that if you diluted a substance, this would "potentize" its ability to cure symptoms, "enhancing" its "spirit-like medicinal powers," and at the same time, as luck would have it, also reducing its side effects. In fact, he went further than this: the more you dilute a substance, the more powerful it becomes at treating the symptoms it would otherwise induce.

Simple dilutions were not enough. Hahnemann decided that the process had to be performed in a very specific way, with an eye on brand identity, or a sense of ritual and occasion, so he devised a process called succussion. With each dilution the glass vessel containing the remedy is shaken by ten firm strikes against "a hard

*At proper high doses, cinchona contains quinine, which can genuinely be used to treat malaria, although most malarial parasites are immune to it now.

but elastic object." For this purpose Hahnemann had a saddle-maker construct a bespoke wooden striking board, covered in leather on one side and stuffed with horsehair. These ten firm strikes are still carried out in homeopathy pill factories today, sometimes by elaborate, specially constructed robots.

Homeopaths have developed a wide range of remedies over the years, and the process of developing them has come to be called, rather grandly, proving (from the German *Prüfung*). A group of volunteers, anywhere from one person to a couple of dozen, come together and take six doses of the remedy being "proved," at a range of dilutions, over the course of two days, keeping a diary of the mental, physical, and emotional sensations, including dreams, experienced over this time. At the end of the proving, the "master prover" will collate the information from the diaries, and this long, unsystematic list of symptoms and dreams from a small number of people will become the "symptom picture" for that remedy, written in a big book and revered, in some cases, for all time. When you go to a homeopath, he or she will try to match your symptoms to the ones caused by a remedy in a proving.

There are obvious problems with this system. For a start, you can't be sure if the experiences the "provers" are having are caused by the substance they're taking or by something entirely unrelated. It might be a "nocebo" effect, the opposite of "placebo," where people feel bad because they're expecting to (I'll bet I could make you feel nauseated right now by telling you some home truths about how your last processed meal was made); it might be a form of group hysteria ("Are there fleas in this sofa?"); one of them might experience a tummy ache that was coming on anyway; or they might all get the same mild cold together; and so on. But homeopaths have been very successful at marketing these "provings" as valid scientific investigations.

Hahnemann professed, and indeed recommended, complete ignorance of the physiological processes going on inside the body;

he treated it as a black box, with medicines going in and effects coming out, and championed only empirical data, the effects of the medicine on symptoms ("The totality of symptoms and circumstances observed in each individual case," he said, "is the one and only indication that can lead us to the choice of the remedy").

This is the polar opposite of the "Medicine only treats the symptoms; we treat and understand the underlying cause" rhetoric of modern alternative therapists. It's also interesting to note, in these times of "natural is good," that Hahnemann said nothing about homeopathy being "natural" and promoted himself as a man of science.

Conventional medicine in Hahnemann's time was obsessed with theory and was hugely proud of basing its practice on a "rational" understanding of anatomy and the workings of the body. Medical doctors in the eighteenth century sneeringly accused homeopaths of "mere empiricism," an overreliance on observations of people getting better. Now the tables are turned; today the medical profession is frequently happy to accept ignorance of the details of mechanism, as long as trial data shows that treatments are effective (we aim to abandon the ones that aren't), whereas homeopaths rely exclusively on their exotic theories and ignore the gigantic swath of negative empirical evidence on their efficacy. It's a small point, perhaps, but these subtle shifts in rhetoric and meaning can be revealing.

THE DILUTION PROBLEM

Before we go any further into homeopathy, and look at whether it actually works or not, there is one central problem we need to get out of the way.

Most people know that homeopathic remedies are diluted to such an extent that there will be no molecules of it left in the dose you get. What you might not know is just how far these remedies are diluted. The typical homeopathic dilution is 30C; this means that the original substance has been diluted by one drop in a hundred, thirty times over. In the "What is homeopathy?" section on the Society of Homeopaths' website, the single largest organization for homeopaths in the U.K. will tell you that "30C contains less than one part per million of the original substance."

"Less than one part per million" is, I would say, something of an understatement: a 30C homeopathic preparation is a dilution of 1 in 100^{30}, or rather 10^{60}, or 1 followed by 60 zeros. To avoid any misunderstandings, this is a dilution of 1 in 1,000,000,000,000,000,000, 000,000,000,000,000,000,000,000,000,000,000,000,000,000, or, to phrase it in the Society of Homeopaths' terms, "one part per million million million million million million million million million million." This is definitely "less than one part per million of the original substance."

For perspective, there are only around 100,000,000,000,000, 000,000,000,000,000,000 molecules of water in an Olympic-size swimming pool. Imagine a sphere of water with a diameter of ninety million miles (the distance from the Earth to the sun). It takes light eight minutes to travel that distance. Picture a sphere of water that size, with one molecule of a substance in it: that's a 30C dilution.*

At a homeopathic dilution of 200C (you can buy much higher dilutions from any homeopathic supplier) the treating substance is diluted more than the total number of atoms in the universe, and by an enormously huge margin. To look at it another way, the universe contains about 3×10^{80} cubic meters of storage space (ideal for starting a family); if it were filled with water and one molecule of active ingredient, this would make for a rather paltry 55C dilution.

*For pedants, it's a 30.89C dilution.

We should remember, though, that the improbability of homeopaths' claims for *how* their pills might work remains fairly inconsequential and is not central to our main observation, which is that they work no better than placebo. We do not know *how* general anesthetics work; but we know that they *do* work, and we use them despite our ignorance of the mechanism. I myself have cut deep into a man's abdomen and rummaged around his intestines in an operating room—heavily supervised, I hasten to add—while he was knocked out by anesthetics, and the gaps in our knowledge regarding their mode of action didn't bother either me or the patient at the time.

Moreover, at the time that homeopathy was first devised by Hahnemann, nobody even knew that these problems existed, because the Italian physicist Amedeo Avogadro and his successors hadn't yet worked out how many molecules there are in a given amount of a given substance, let alone how many atoms there are in the universe. We didn't even really know what atoms were.

How have homeopaths dealt with the arrival of this new knowledge? By saying that the absent molecules are irrelevant, because "water has a memory." This sounds feasible if you think of a bath or a test tube full of water. But if you think, at the most basic level, about the scale of these objects, a tiny water molecule isn't going to be deformed by an enormous arnica molecule and be left with a "suggestive dent," which is how many homeopaths seem to picture the process. A pea-size lump of putty cannot take an impression of the surface of your sofa.

Physicists have studied the structure of water very intensively for many decades, and while it is true that water molecules will form structures around a molecule dissolved in them at room temperature, the everyday random motion of water molecules means that these structures are very short-lived, with lifetimes measured in picoseconds, or even less. This is a very restrictive shelf life.

Homeopaths will sometimes pull out anomalous results from physics experiments and suggest that these prove the efficacy of homeopathy. They have fascinating flaws, which can be read about elsewhere (frequently the homeopathic substance, which is found on hugely sensitive lab tests to be subtly different from a nonhomeopathic dilution, has been prepared in a completely different way, from different stock ingredients, which is then detected by exquisitely sensitive lab equipment). As a ready shorthand, it's also worth noting that the American magician and debunker James Randi has offered a one-million-dollar prize to anyone demonstrating "anomalous claims" under laboratory conditions, and has specifically stated that anyone could win it by reliably distinguishing a homeopathic preparation from a nonhomeopathic one using any method they wish. This one-million-dollar bounty remains unclaimed.

Even if taken at face value, the "memory of water" claim has large conceptual holes, and most of them you can work out for yourself. If water has a memory, as homeopaths claim, and a 1 in 10^{60} dilution is fine, then by now all water must surely be a health-giving homeopathic dilution of all the molecules in the world. Water has been sloshing around the globe for a very long time after all, and the water in my very body as I sit here typing away in London has already been through plenty of other people's bodies before mine. Maybe some of the water molecules sitting in my fingers as I type this sentence are currently in your eyeball. Maybe some of the water molecules fleshing out my neurons as I decide whether to write "wee" or "urine" in this sentence are now in the bladder of the queen of England (God bless her). Water is a great leveler; it gets about. Just look at clouds.

How does a water molecule know to forget every other molecule it's seen before? How does it know to treat my bruise with its memory of arnica, rather than a memory of Isaac Asimov's feces? I wrote this in the newspaper once, and a homeopath complained to the Press Complaints Commission. It's not about the dilution, he said; it's the succussion. You have to bang the flask of water briskly ten times on a leather and horsehair surface, and that's what makes the water remember a molecule. Because I did not mention this, he explained, *I had deliberately made homeopaths sound stupid.* This is another universe of foolishness.

And for all homeopaths' talk about the "memory of water," we should remember that what you actually take, in general, is a little sugar pill, not a teaspoon of homeopathically diluted water, so they should start thinking about the memory of sugar too. The memory of sugar, which is remembering something that was being remembered by water (after a dilution greater than the number of atoms in the universe) but then got passed on to the sugar as it dried. I'm trying to be clear, because I don't want any more complaints.

Once this sugar, which has remembered something the water was remembering, gets into your body, it must have some kind of effect. What would that be? Nobody knows, but you need to take the pills regularly, apparently, in a dosing regime that is suspiciously similar to that for medical drugs (which are given at intervals spaced according to how fast they are broken down and excreted by your body).

I DEMAND A FAIR TRIAL

These theoretical improbabilities are interesting, but they're not going to win you any arguments: Sir John Forbes, physician to Queen Victoria, pointed out the dilution problem in the nineteenth century, and 150 years later the discussion has not moved on. The real question with homeopathy is very simple: Does it work? In fact, how do we know if *any* given treatment is working?

Symptoms are a very subjective thing, so almost every conceivable way of establishing the benefits of any treatment must start with the individual and his or her experience, building from there. Let's imagine we're talking—maybe even arguing—with someone who thinks that homeopathy works, someone who feels it is a positive experience, and who feels he gets better, quicker, with home-opathy. They would say: "All I know is, I feel as if it works. I get better when I take homeopathy." It seems obvious to them, and to an extent it is. This statement's power, and its flaws, lie in its sim-plicity. Whatever happens, the statement stands as true.

But you could pop up and say: "Well, perhaps that was the placebo effect." Because the placebo effect is far more complex and interesting than most people suspect, going way beyond a mere sugar pill; it's about the whole cultural experience of a treatment, your expectations beforehand, the consultation process you go through while receiving the treatment, and much more.

We know that two sugar pills are a more effective treatment than one sugar pill, for example, and we know that saltwater in-jections are a more effective treatment for pain than sugar pills, not because saltwater injections have any biological action on the body, but because an injection feels like a more dramatic interven-tion. We know that the color of pills, their packaging, how much you pay for them, and even the beliefs of the people handing the pills over are all important factors. We know that placebo opera-tions can be effective for knee pain and even for chest pain. The placebo effect works on animals and children. It is highly potent, and very sneaky, and you won't know the half of it until you read the placebo chapter in this book.

So when our homeopathy fan says that homeopathic treat-ment makes them feel better, we might reply: "I accept that, but perhaps your improvement is because of the placebo effect," and they cannot answer no, because they have *no possible way of know-ing* whether they got better through the placebo effect or not. They cannot tell. The most they can do is restate, in response to your query, their original statement: "All I know is, I feel as if it works. I get better when I take homeopathy."

Next, you might say: "OK, I accept that, but perhaps, also, you feel you're getting better because of 'regression to the mean.'" This is just one of the many "cognitive illusions" described in this book, the basic flaws in our reasoning apparatus that lead us to see patterns and connections in the world around us, when closer inspection reveals that in fact, there are none.

"Regression to the mean" is basically another phrase for the phenomenon whereby, as alternative therapists like to say, all things have a natural cycle. Let's say you have back pain. It comes and goes. You have good days and bad days, good weeks and bad weeks. When it's at its very worst, it's going to get better, because that's the way things are with your back pain.

Similarly, many illnesses have what is called a natural history: they are bad, and then they get better. As Voltaire said, "The art of medicine consists in amusing the patient while nature cures the disease." Let's say you have a cold. It's going to get better after a few days, but at the moment you feel miserable. It's quite natural that when your symptoms are at their very worst, you will do things to try to get better. You might take a homeopathic remedy. You might sacrifice a goat and dangle its entrails around your neck. You might bully your physician into giving you antibiotics. (I've listed these in order of increasing ridiculousness.)

Then, when you get better—as you surely will from a cold— you will naturally assume that whatever you did when your symptoms were at their worst must be the reason for your recovery. *Post hoc, ergo propter hoc*, and all that. Every time you get a cold from now on, you'll be back at your physician, hassling her for antibiotics, and she'll be saying, "Look, I don't think this is a very good idea," but you'll insist, because they worked last time, and community antibiotic resistance will increase, and ultimately old ladies die from multiple-drug-resistant bacteria, because of this kind of irrationality, but that's another story.*

---

*Physicians sometimes prescribe antibiotics to demanding patients in exasperation, even though they are ineffective in treating a viral cold, but much research suggests that this is counterproductive, even as a time-saver. In one study, prescribing antibiotics rather than giving advice on self-management for sore throat resulted in an increased overall

You can look at regression to the mean more mathematically, if you prefer. On *Card Sharks*, when the host puts a three on the board, the audience all shout, "Higher!" because they know the odds are that the next card is going to be higher than a three. "Do you want to go higher or lower than a jack? Higher? Higher?" "Lower!"

An even more extreme version of regression to the mean is what is known as the *Sports Illustrated* jinx. Whenever a sports-man appears on the cover of *Sports Illustrated*, goes the story, he is soon to fall from grace. But to get on the cover of the magazine, you have to be at the absolute top of your game, one of the best sportsmen in the world, and to be the best in that week, you're probably also having an unusual run of luck. Luck, or "noise," generally passes; it "regresses to the mean" by itself, as happens with throws of a die. If you fail to understand that, you start look-ing for another cause for that regression, and you find . . . the *Sports Illustrated* jinx.

Homeopaths increase the odds of a perceived success in their treatments even further by talking about aggravations, explaining that sometimes the correct remedy can make symptoms get worse before they get better, and claiming that this is part of the treat-ment process. Similarly, people flogging detox will often say that their remedies might make you feel worse at first, as the toxins are extruded from your body; under the terms of these promises, liter-ally anything that happens to you after a treatment is proof of the therapist's clinical acumen and prescribing skill.

So we could go back to our homeopathy fan and say: "You feel you get better, I accept that. But perhaps it is because of re-

---

workload through repeat attendance. It was calculated that if a doctor prescribed antibiotics for sore throat to one hundred fewer patients each year, thirty-three fewer would believe that antibiotics were effective, twenty-five fewer would intend to consult with the problem in the future, and ten fewer would come back within the next year. If you were an alternative therapist, or a drug salesman, you could turn those figures on their head and look at how to drum up more trade, not less.

gression to the mean, or simply the natural history of the disease."
Again, he cannot say no (or at least not with any meaning—
he might say it in a tantrum), because he has no possible way of
knowing whether he was going to get better anyway on the occasions when he apparently got better after seeing a homeopath.
Regression to the mean might well be the true explanation for
his return to health. He simply cannot tell. He can only restate,
again, his original statement: "All I know is, I feel as if it works. I
get better when I take homeopathy."

That may be as far as he wants to go. But when someone goes
further and says, "Homeopathy works," or mutters about "science,"
then that's a problem. We cannot simply decide such things on
the basis of one individual's experiences, for the reasons described
above: they might be mistaking the placebo effect for a real effect
or mistaking a chance finding for a real one. Even if we had one
genuine, unambiguous, and astonishing case of a person's getting
better from terminal cancer, we'd still be careful about using that
one person's experience, because sometimes, entirely by chance,
miracles really do happen. Sometimes, but not very often.

Over the course of many years, a team of Australian oncologists
followed 2,337 terminal cancer patients in palliative care. They
died, on average, after five months. But around 1 percent of them
were still alive after five years. In January 2006 this study was reported in *The Independent* newspaper in the U.K., bafflingly, as:

"MIRACLE" CURES SHOWN TO WORK
Doctors have found statistical evidence that alternative
treatments such as special diets, herbal potions and faith
healing can cure apparently terminal illness, but they remain unsure about the reasons.

But the point of the study was specifically *not* that there are miracle cures (it didn't look at any such treatments; that was an inven-

tion by the newspaper). Instead it showed something much more interesting: that amazing things simply happen sometimes; people can survive, despite all the odds, for no apparent reason. As the researchers made clear in their own description, claims for miracle cures should be treated with caution, because "miracles" occur routinely, in 1 percent of cases by their definition, and *without* any specific intervention. The lesson of this paper is that we cannot reason from one individual's experience or even that of a handful, selected out to make a point.

So how do we move on? The answer is that we take lots of individuals, a sample of patients who represent the people we hope to treat, with all of their individual experiences, and count them all up. This is clinical academic medical research, in a nutshell, and there's really nothing more to it than that: no mystery, no "different paradigm," no smoke and mirrors. It's an entirely transparent process, and this one idea has probably saved more lives, on a more spectacular scale, than any other idea you will come across this year.

It is also not a new idea. The first trial appears in the Old Testament, and interestingly, although nutritionism has only recently become what we might call the bullshit du jour, it was about food. Daniel was arguing with King Nebuchadnezzar's chief eunuch over the Judaean captives' rations. Their diet was rich food and wine, but Daniel wanted his own soldiers to be given only vegetables. The eunuch was worried that they would become worse soldiers if they didn't eat their rich meals, and that whatever could be done to a eunuch to make his life worse might be done to him. Daniel, on the other hand, was willing to compromise, so he suggested the first ever clinical trial:

> And Daniel said unto the guard . . . "Submit us to this test for ten days. Give us only vegetables to eat and water to drink; then compare our looks with those of the young

men who have lived on the food assigned by the king and
be guided in your treatment of us by what you see."

The guard listened to what they said and tested them
for ten days. At the end of ten days they looked healthier
and were better nourished than all the young men who
had lived on the food assigned them by the king. So the
guard took away the assignment of food and the wine they
were to drink and gave them only the vegetables.

—Daniel 1:1–16

To an extent, that's all there is to it; there's nothing particularly
mysterious about a trial, and if we wanted to see whether homeopathy pills work, we could do a very similar trial. Let's flesh it out.
We would take, say, two hundred people going to a homeopathy
clinic, divide them randomly into two groups, and let them go
through the whole process of seeing the homeopath, being diag-
nosed, and getting their prescription for whatever the homeopath
wants to give them. But at the last minute, without their knowledge, we would switch half of the patients' homeopathic sugar pills,
giving them dud sugar pills, that have not been magically "poten-
tized" by homeopathy. Then, at an appropriate time later, we could
measure how many in each group got better.

Speaking with homeopaths, I have encountered a great deal
of angst about the idea of measuring, as if this were somehow not
a transparent process, as if it were forcing a square peg into a round
hole, because "measuring" sounds scientific and mathematical. We
should pause for just a moment and think about this clearly. Measuring involves no mystery and no special devices. We ask people
if they feel better and count up the answers.

In a trial—or sometimes routinely in outpatients' clinic—we
might ask people to measure their knee pain on a scale of one
to ten every day, in a diary. Or to count up the number of pain-
free days in a week. Or to measure the effect their fatigue has had

on their lives that week: how many days they've been able to get out of the house, how far they've been able to walk, how much housework they've been able to do. You can ask about any number of very simple, transparent, and often quite subjective things, because the business of medicine is improving lives and ameliorating distress.

We might dress the process up a bit, to standardize it, and allow our results to be compared more easily with other research (which is a good thing, as it helps us get a broader understanding of a condition and its treatment). We might use the General Health Questionnaire, for example, because it's a standardized "tool," but for all the bluster, the GHQ-12, as it is known, is just a simple list of questions about your life and your symptoms.

If antiauthoritarian rhetoric is your thing, then bear this in mind: perpetrating a placebo-controlled trial of an accepted treatment—whether it's an alternative therapy or any form of medicine—is an inherently subversive act. You undermine false certainty, and you deprive doctors, patients, and therapists of treatments that previously pleased them.

There is a long history of upset being caused by trials, in medicine as much as anywhere, and all kinds of people will mount all kinds of defenses against them. Archie Cochrane, one of the grandfathers of evidence-based medicine, once amusingly described how different groups of surgeons were each earnestly contending that their treatment for cancer was the most effective; it was transparently obvious to them all that their own treatment was the best. Cochrane went so far as to bring a collection of them together in a room, so that they could witness one another's dogged but conflicting certainty, in his efforts to persuade them of the need for trials. Judges, similarly, can be highly resistant to the notion of trialing different forms of sentence for heroin users, even though there is no evidence to say which kind of sentence (custodial, compulsory drug treatment, and so on) is best. They believe that they can divine the most appropriate sentence in each individual case, without the need for experimental data. These are recent battles, and they are in no sense unique to the world of homeopathy.

So, we take our group of people coming out of a homeopathy clinic, we switch half their pills for placebo pills, and we measure who gets better. That's a placebo-controlled trial of homeopathy pills, and this is not a hypothetical discussion; these trials have been done on homeopathy, and it seems that overall, homeopathy does no better than placebo.

And yet you will have heard homeopaths say that there are positive trials in homeopathy; you may even have seen specific ones quoted. What's going on here? The answer is fascinating, and takes us right to the heart of evidence-based medicine. There are *some* trials that find homeopathy performs better than placebo, but only some, and they are, in general, trials with "methodological flaws." This sounds technical, but all it means is that there are problems in the way the trials were performed, and those problems are so great that they mean the trials are less "fair tests" of a treatment.

The alternative therapy literature is certainly riddled with incompetence, but flaws in trials are actually very common throughout medicine. In fact, it would be fair to say that all research has some flaws, simply because every trial will involve a compromise between what would be ideal and what is practical or cheap. (The literature from complementary and alternative medicine—CAM—often fails badly at the stage of interpretation; medics sometimes know if they're quoting duff papers and describe the flaws, whereas homeopaths tend to be uncritical of anything positive.)

That is why it's important that research is always published, in full, with its methods and results available for scrutiny. This is a recurring theme in this book, and it's important, because when people make claims based upon their research, we need to be able to decide for ourselves how big the "methodological flaws" were, and come to our own judgment about whether the results are reliable, whether theirs was a "fair test." The things that stop a trial from being fair are, once you know about them, blindingly obvious.

BLINDING

One important feature of a good trial is that neither the experimenters nor the patients know if they got the homeopathy sugar pill or the simple placebo sugar pill, because we want to be sure that any difference we measure is the result of the difference between the pills and not of people's expectations or biases. If the researchers knew which of their beloved patients were having the real and which the placebo pills, they might give the game away or it might change their assessment of the patient—consciously or unconsciously.

Let's say I'm doing a study on a medical pill designed to reduce high blood pressure. I know which of my patients are having the expensive new blood pressure pill and which are having the placebo. One of the people on the swanky new blood pressure pills comes in and has a blood pressure reading that is way off the scale, much higher than I would have expected, especially since he's on this expensive new drug. So I recheck his blood pressure, "just to make sure I didn't make a mistake." The next result is more normal, so I write that one down and ignore the high one.

Blood pressure readings are an inexact technique, like ECG interpretation, X-ray interpretation, pain scores, and many other measurements that are routinely used in clinical trials. I go for lunch, entirely unaware that I am calmly and quietly polluting the data, destroying the study, producing inaccurate evidence, and therefore, ultimately, killing people (because our greatest mistake would be to forget that data is used for serious decisions in the very real world, and bad information causes suffering and death).

There are several good examples from recent medical history where a failure to ensure adequate blinding, as it is called, has resulted in the entire medical profession's being mistaken about which was the better treatment. We had no way of knowing whether keyhole surgery was better than open surgery, for example, until a group of surgeons from Sheffield came along and did a very theatrical trial, in which bandages and decorative fake blood squirts were used, to make sure that nobody could tell which type of operation anyone had received.

Some of the biggest figures in evidence-based medicine got together and did a review of blinding in all kinds of trials of medical drugs and found that trials with inadequate blinding exaggerated the benefits of the treatments being studied by 17 percent. Blinding is not some obscure piece of nitpicking, idiosyncratic to pedants like me, used to attack alternative therapies.

Closer to home for homeopathy, a review of trials of acupuncture for back pain showed that the studies that were properly blinded showed a tiny benefit for acupuncture, which was not "statistically significant" (we'll come back to what that means later). Meanwhile, the trials that were not blinded—the ones in which the patients knew whether they were in the treatment group or not—showed a massive, statistically significant benefit for acupuncture. (The placebo control for acupuncture, in case you're wondering, is sham acupuncture, with fake needles or needles in the "wrong" places, although an amusing complication is that sometimes one school of acupuncturists will claim that another school's sham needle locations are actually their genuine ones.)

So, as we can see, blinding is important, and not every trial is necessarily any good. You can't just say, "Here's a trial that shows this treatment works," because there are good trials, or "fair tests," and there are bad trials. When doctors and scientists say that a

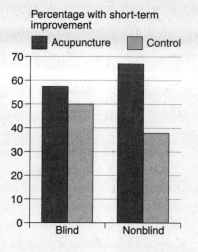

Percentage with short-term improvement

study was methodologically flawed and unreliable, it's not because they're being mean, or trying to maintain the "hegemony," or to keep the backhanders coming from the pharmaceutical industry; it's because the study was poorly performed—it costs nothing to blind properly—and simply wasn't a fair test.

RANDOMIZATION

Let's take this out of the theoretical, and look at some of the trials that homeopaths quote to support their practice. I've got in front of me, a standard review of trials for homeopathic arnica by homeopathist professor Edzard Ernst,* which we can go through for examples. We should be absolutely clear that the inadequacies here are not unique, I do not imply intent to deceive, and I am not being mean. What we are doing is simply what medics and academics do when they appraise evidence.

*Ernst has been awarded the first chair in complementary medicine at the University of Exeter.

So, Hildebrandt et al. (as they say in academia) looked at forty-two women taking homeopathic arnica for delayed-onset muscle soreness and found it performed better than placebo. At first glance this seems to be a pretty plausible study, but if you look closer, you can see there was no randomization described. Randomization is another basic concept in clinical trials. We randomly assign patients to the placebo sugar pill group or the homeopathy sugar pill group, because otherwise there is a risk that the doctor or homeopath—consciously or unconsciously—will put patients who they think might do well into the homeopathy group and the no-hopers into the placebo group, thus rigging the results.

Randomization is not a new idea. It was first proposed in the seventeenth century by Jan Baptista van Helmont, a Belgian radical who challenged the academics of his day to test their treatments like bloodletting and purging (based on "theory") against his own, which he said were based more on clinical experience: "Let us take out of the hospitals, out of the Camps, or from elsewhere, two hundred, or five hundred poor People, that have Fevers, Pleurisies, etc. Let us divide them into half, let us cast lots, that one half of them may fall to my share, and the other to yours . . . We shall see how many funerals both of us shall have."

It's rare to find an experimenter so careless that he's not randomized the patients at all, even in the world of CAM. But it's surprisingly common to find trials in which the method of randomization is inadequate: they look plausible at first glance, but on closer examination we can see that the experimenters have simply gone through a kind of theater, as if they were randomizing the patients but still leaving room for them to influence, consciously or unconsciously, which group each patient goes into.

In some inept trials, in all areas of medicine, patients are randomized into the treatment or placebo group by the order in which they are recruited into the study: the first patient in gets the real treatment, the second gets the placebo, the third the real treatment, the fourth the placebo, and so on. This sounds fair enough, but in fact, it's a glaring hole that opens your trial up to possible systematic bias.

Let's imagine there is a patient who the homeopath believes to be a no-hoper, a "heart-sink" patient who'll never really get better, no matter what treatment he or she gets, and the next place available on the study is for someone going into the "homeopathy" arm of the trial. It's not inconceivable that the homeopath might just decide—again, consciously or unconsciously—that this particular patient "probably wouldn't really be interested" in the trial. But if, on the other hand, this no-hoper patient had come into clinic at a time when the next place on the trial was for the placebo group, the recruiting clinician might have felt a lot more optimistic about signing him up.

The same goes for all the other inadequate methods of randomization: by last digit of date of birth, by date seen in clinic, and so on. There are even studies that claim to randomize patients by tossing a coin, but forgive me (and the entire evidence-based medicine community) for worrying that tossing a coin leaves itself just a little bit too open to manipulation. Best of three, and all that. Sorry, I meant best of five. Oh, I didn't really see that one: it fell on the floor.

There are plenty of genuinely fair methods of randomization, and although they require a bit of effort, they come at no extra financial cost. The classic is to make people call a special telephone number, to where someone is sitting with a computerized randomization program (and the experimenter doesn't even do that until the patient is fully signed up and committed to the study). This is probably the most popular method among meticulous researchers, who are keen to ensure they are doing a "fair test," simply because you'd have to be an out-and-out charlatan to mess it up, and you'd have to work pretty hard at the charlatanry too.

Does randomization matter? As with blinding, people have studied the effect of randomization in huge reviews of large numbers of trials and found that the ones with dodgy methods of randomization overestimate treatment effects by 41 percent. In reality, the biggest problem with poor-quality trials is not that they've used an inadequate method of randomization; it's that they don't tell you *how* they randomized the patients at all. This is a classic warning sign and often means the trial has been performed badly. Again, I do not speak from prejudice: trials with unclear methods of randomization overstate treatment effects by 30 percent, almost as much as the trials with openly rubbish methods of randomization.

In fact, as a general rule it's always worth worrying when people don't give you sufficient details about their methods and results. As it happens (I promise I'll stop this soon), there have been two landmark studies on whether inadequate information in academic articles is associated with dodgy, overly flattering results, and yes, studies that don't report their methods fully do overstate the benefits of the treatments, by around 25 percent. Transparency and detail are everything in science. Hildebrandt et al., through no fault of their own, happened to be the peg for this discussion on randomization (and I am grateful to them for it). They might well have randomized their patients. They might well have done so adequately. But they did not report on it.

Let's go back to the eight studies in Ernst's review article on homeopathic arnica, which we chose pretty arbitrarily, because they demonstrate a phenomenon that we see over and over again with complementary and alternative medicine (CAM) studies: most of the trials were hopelessly methodologically flawed and showed positive results for homeopathy, whereas the couple of decent studies—the most "fair tests"—showed homeopathy to perform no better than placebo.*

*So, Pinsent performed a double-blind, placebo-controlled study of fifty-nine people having oral surgery. The group receiving homeopathic arnica experienced significantly less pain than the group getting placebo. What you don't tend to read in the arnica pub-

So now you can see, I would hope, that when doctors say a piece of research is "unreliable," that's not necessarily a scam, where academics deliberately exclude a poorly performed study that flatters homeopathy, or any other kind of paper, from a systematic review of the literature, and it's not through a personal or moral bias: it's for the simple reason that if a study is no good, if it is not a "fair test" of the treatments, then it might give unreliable results, and so it should be regarded with great caution.

There is a moral and financial issue here too: randomizing your patients properly doesn't cost money. Blinding your patients to whether they had the active treatment or the placebo doesn't cost money. Overall, doing research robustly and fairly does not necessarily require more money; it simply requires that you think before you start. The only people to blame for the flaws in these studies are the people who performed them. In some cases they will be people who turn their backs on the scientific method as a

---

licity material is that forty-one subjects dropped out of this study. That makes it a fairly rubbish study. It's been shown that patients who drop out of studies are less likely to have taken their tablets properly, more likely to have had side effects, less likely to have got better, and so on. I am not skeptical about this study because it offends my prejudices but because of the high dropout rate. The missing patients might have been lost to followup because they are dead, for example. Ignoring dropouts tends to exaggerate the benefits of the treatment being tested, and a high dropout rate is always a warning sign.

The study by Gibson et al. did not mention randomization, nor did it deign to mention the dose of the homeopathic remedy, or the frequency with which it was given. It's not easy to take studies very seriously when they are this thin.

There was a study by Campbell that had thirteen subjects in it (meaning a tiny handful of patients in both the homeopathy and the placebo groups). It found that homeopathy performed better than placebo (in this teeny-tiny sample of subjects), but didn't check whether the results were statistically significant or merely chance findings.

Last, Savage et al. did a study with a mere ten patients, finding that homeopathy was better than placebo, but they too did no statistical analysis of their results.

These are the kinds of papers that homeopaths claim as evidence to support their case, evidence that they claim is deceitfully ignored by the medical profession. All these studies favored homeopathy. All deserve to be ignored, for the simple reason that each was not a fair test of homeopathy, simply on account of these methodological flaws.

I could go on, through a hundred homeopathy trials, but it's painful enough already.

"flawed paradigm," and yet it seems their great new paradigm is simply "unfair tests."

These patterns are reflected throughout the alternative therapy literature. In general, the studies that are flawed tend to be the ones that favor homeopathy, or any other alternative therapy, and the well-performed studies, in which every controllable source of bias and error is excluded, tend to show that the treatments are no better than placebo.

This phenomenon has been carefully studied, and there is an almost linear relationship between the methodological quality of a homeopathy trial and the result it gives. The worse the study— which is to say, the less it is a "fair test"—the more likely it is to find that homeopathy is better than placebo. Academics conventionally measure the quality of a study using standardized tools like the Jadad score, a seven-point tick list that includes things we've been talking about, like "Did they describe the method of randomization?" and "Was plenty of numerical information provided?"

This graph, from Ernst's paper, shows what happens when you plot Jadad score against result in homeopathy trials. Toward the top left, you can see rubbish trials with huge design flaws that triumphantly find that homeopathy is much, much better than placebo. Toward the bottom right, you can see that as the Jadad score tends toward the top mark of 5, as the trials become more of a "fair test," the line tends toward showing that homeopathy performs no better than placebo. There is, however, a mystery in this graph, an oddity, and the makings of a whodunit. That little dot on the right-hand edge of the graph, representing the ten best-quality trials, with the highest Jadad scores, stands clearly outside the trend of all the others. This is an anomalous finding; suddenly, only at that end of the graph, there are some good-quality trials bucking the trend and showing that homeopathy is better than placebo.

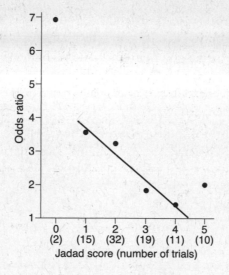

Jadad score (number of trials)

What's going on there? I can tell you what I think: some of the papers making up that spot are rigged. I don't know which ones, how it happened, or who did it, in which of the ten papers, but that's what I think. Academics often have to couch strong criticism in diplomatic language. Here is Professor Ernst, the man who made that graph, discussing the eyebrow-raising outlier. You might decode his political diplomacy and conclude that he thinks there's been a fix too.

> There may be several hypotheses to explain this phenomenon. Scientists who insist that homeopathic remedies are in every way identical to placebos might favor the following. The correlation provided by the four data points (Jadad score 1–4) roughly reflects the truth. Extrapolation of this correlation would lead them to expect that those trials with the least room for bias (Jadad score = 5) show homeopathic remedies are pure placebos. The fact, however, that the average result of the 10 trials scoring 5 points

on the Jadad score contradicts this notion, is consistent
with the hypothesis that some (by no means all) methodologically astute and highly convinced homeopaths have
published results that look convincing but are, in fact, not
credible.

But this is a curiosity and an aside. In the bigger picture it doesn't
matter, because overall, even including these suspicious studies,
the meta-analyses still show, overall, that homeopathy is no better
than placebo.

Meta-analyses?

META-ANALYSIS

This will be our last big idea for a while, and this is one that has
saved the lives of more people than you will ever meet. A meta-
analysis is a very simple thing to do, in some respects: you just collect all the results from all the trials on a given subject, bung them
into one big spreadsheet, and do the math on that, instead of relying
on your own gestalt intuition about all the results from each of your
little trials. It's particularly useful when there have been lots of trials, each too small to give a conclusive answer, but all looking at
the same topic.

So if there are, say, ten randomized, placebo-controlled trials
looking at whether asthma symptoms get better with homeopathy,
each of which has a paltry forty patients, you could put them all
into one meta-analysis and effectively (in some respects) have a
four-hundred-person trial to work with.

In some very famous cases—at least, famous in the world of
academic medicine—meta-analyses have shown that a treatment
previously believed to be ineffective is in fact rather good, but
because the trials that had been done were each too small, indi-

vidually, to detect the real benefit, nobody had been able to spot it.

As I said, information alone can be lifesaving, and one of the greatest institutional innovations of the past thirty years is undoubtedly the Cochrane Collaboration, an international not-for-profit organization of academics that produces systematic summaries of the research literature on health care research, including meta-analyses.

The logo of the Cochrane Collaboration features a simplified blobbogram, a graph of the results from a landmark meta-analysis that looked at an intervention given to pregnant mothers. When women give birth prematurely, as you might expect, the babies are more likely to suffer and die. Some doctors in New Zealand had the idea that giving a short, cheap course of a steroid might help improve outcomes, and seven trials testing this idea were done between 1972 and 1981. Two of them showed some benefit from the steroids, but the remaining five failed to detect any benefit, and because of this, the idea didn't catch on.

**THE COCHRANE
COLLABORATION®**

Eight years later, in 1989, a meta-analysis was done by pooling all this trial data. If you look at the blobbogram in the logo on the previous page, you can see what happened. Each horizontal line represents a single study: if the line is over to the left, it means the steroids were better than placebo, and if it is over to the right, it means the steroids were worse. If the horizontal line for a trial touches the big vertical nil effect line going down the middle, then the trial showed no clear difference either way. One last thing: the longer a horizontal line is, the less certain the outcome of the study was.

Looking at the blobbogram, we can see that there are lots of not-very-certain studies, long horizontal lines, mostly touching the central vertical line of no effect; but they're all a bit over to the left, so they all seem to suggest that steroids *might* be beneficial, even if each study itself is not statistically significant.

The diamond at the bottom shows the pooled answer: that there is in fact very strong evidence indeed for steroids reducing the risk—by 30 to 50 percent—of babies dying from the complications of immaturity. We should always remember the human cost of these abstract numbers: babies died unnecessarily because they were deprived of this lifesaving treatment for a decade. They died *even when there was enough information available to know what would save them*, because that information had not been synthesized together, and analyzed systematically, in a meta-analysis.

Back to homeopathy (you can see why I find it trivial now). A landmark meta-analysis was published recently in *The Lancet*. It was accompanied by an editorial titled "The End of Homeopathy?" Shang et al. did a very thorough meta-analysis of a vast number of homeopathy trials, and they found, overall, adding them all up, that homeopathy performs no better than placebo.

The homeopaths were up in arms. If you mention this meta-analysis, they will try to tell you that it was a fix. What Shang et al. did, essentially, like all the previous negative meta-analyses of homeopathy, was to exclude the poorer-quality trials from their analysis.

Homeopaths like to pick out the trials that give them the answer that they want to hear and ignore the rest, a practice called cherry-picking. But you can also cherry-pick your favorite meta-analyses or misrepresent them. Shang et al. were only the latest in a long string of meta-analyses to show that homeopathy performs no better than placebo. What is truly amazing to me is that despite the negative results of these meta-analyses, homeopaths have continued—right to the top of the profession—to claim that these same meta-analyses *support* the use of homeopathy. They do this by quoting only the result for *all* trials included in each meta-analysis. This figure includes all of the poorer-quality trials. The most reliable figure, you now know, is for the restricted pool of the most "fair tests," and when you look at those, homeopathy performs no better than placebo. If this fascinates you (and I would be very surprised), then I am currently producing a summary with some colleagues, and you will soon be able to find it online at badscience .net, in all its glorious detail, explaining the results of the various meta-analyses performed on homeopathy.

Clinicians, pundits, and researchers all like to say things like "There is a need for more research," because it sounds forward-thinking and open-minded. In fact, that's not always the case, and it's a little-known fact that this very phrase has been effectively banned from the *British Medical Journal* for many years, on the ground that it adds nothing; you may say what research is missing, on whom, how, measuring what, and why you want to do it, but the hand-waving, superficially open-minded call for "more research" is meaningless and unhelpful.

There have been more than a hundred randomized placebo-controlled trials of homeopathy, and the time has come to stop. Homeopathy pills work no better than placebo pills; we know that much. But there is room for more interesting research. People do experience that homeopathy is positive for them, but the action is likely to be in the whole process of going to see a homeopath, of being listened to, having some kind of explanation for your symptoms, and all the other collateral benefits of old-fashioned, paternalistic, reassuring medicine. (Oh, and regression to the mean.)

So we should measure that, and here is the final superb lesson in evidence-based medicine that homeopathy can teach us: sometimes you need to be imaginative about what kinds of research you do, compromise, and be driven by the questions that need answering, rather than by the tools available to you.

It is very common for researchers to research the things that interest them, in all areas of medicine, but they can be interested in quite different things from patients. One study actually thought to ask people with osteoarthritis of the knee what kind of research they wanted to be carried out, and the responses were fascinating: they wanted rigorous real-world evaluations of the benefits from physiotherapy and surgery, from educational and coping strategy interventions, and other pragmatic things. They didn't want yet another trial comparing one pill with another or with placebo.

In the case of homeopathy, similarly, homeopaths want to believe that the power is in the pill, rather than in the whole process of going to visit a homeopath, having a chat, and so on. It is crucially important to their professional identity. But I believe that going to see a homeopath is probably a helpful intervention, in some cases, for some people, even if the pills are just placebos. I think patients would agree, and I think it would be an interesting thing to measure. It would be easy, and you would do something called a pragmatic waiting list–controlled trial.

You take two hundred patients, say, all suitable for homeopathic treatment, currently in a doctor's clinic, and all willing to be referred on for homeopathy, then split them randomly into two groups of one hundred. One group gets treated by a homeopath as normal, pills, consultation, smoke, and voodoo, on top of whatever other treatment they are having, same as in the real world. The other group just sits on the homeopathy waiting list, so they get "treatment as usual," whether that is "neglect," "family doctor treatment," or whatever, but they get no homeopathy. Then you measure outcomes and compare who gets better the most.

You could argue that it would be a trivial positive finding, and that it's obvious the homeopathy group would do better; but it's the only piece of research really waiting to be done. This is a "pragmatic trial." The groups aren't blinded, but they couldn't possibly be in this kind of trial, and sometimes we have to accept compromises in experimental methodology. It would be a legitimate use of public money (or perhaps money from Boiron, the homeopathic pill company valued at five hundred million dollars), but there's nothing to stop homeopaths from just cracking on and doing it for themselves, because despite the homeopaths' fantasies, born out of a lack of knowledge, that research is difficult, magical, and expensive, in fact, such a trial would be very cheap to conduct.

But it's not really money that's missing from the alternative therapy research community working with the ideas of this billion-dollar industry; it's knowledge of evidence-based medicine and expertise in how to do a trial. Their literature and debates drip with ignorance and vitriolic anger at anyone who dares to appraise the trials. Their university courses, as far as they ever even dare to admit what they teach on them (it's all suspiciously hidden away), seem to skirt around such explosive and threatening questions. I've suggested in various places, including at British academic conferences, that the single thing that would most improve the quality of evidence in CAM would be funding for a simple, evidence-based medicine hotline that anyone thinking about running a trial in a clinic could phone up and get advice on how to do it properly, to avoid wasting effort on an "unfair test" that would rightly be regarded with contempt by all outsiders.

In my pipe dream (I'm completely serious, if you've got the money) you'd need a handout, maybe a short course that people did to cover the basics, so they weren't asking stupid questions, and phone support. In the meantime, if you're a sensible homeopath and you want to do a pragmatic, "waiting-list-controlled trial" as I described above, you could maybe try the badscience website forums, where there are people who might be able to give some pointers (among the childish fighters and trolls . . .).

But would the homeopaths buy it? I think it would offend their sense of professionalism. You often see homeopaths trying to nuance their way through this tricky area, and they can't quite make their minds up. Here, for example, is a Radio 4 interview, archived in full online, in which Dr. Elizabeth Thompson (consultant homeopathic physician and honorary senior lecturer at the Department of Palliative Medicine at the University of Bristol) has a go.

She starts off with some sensible stuff: homeopathy does work, but through nonspecific effects, the cultural meaning of the process, the therapeutic relationship, it's not about the pills, and so on. She practically comes out and says that homeopathy is all about cultural meaning and the placebo effect. "People have wanted to say homeopathy is like a pharmaceutical compound," she says, "and it isn't, it is a complex intervention."

Then the interviewer asks: "What would you say to people who go along to their high street pharmacy, where you can buy homeopathic remedies, they have hay fever and they pick out a hay-fever remedy, I mean presumably that's not the way it works?" There is a moment of tension. Forgive me, Dr. Thompson, but I felt you didn't want to say that the pills work, as pills, in isolation, when you buy them in a shop; apart from anything else, you'd already said that they don't.

But she doesn't want to break ranks and say the pills don't work, either. I'm holding my breath. How will she do it? Is there a linguistic structure complex enough, passive enough, to negotiate through this? If there is, Dr. Thompson doesn't find it: "They might flick through and they might just be spot-on . . . [but] you've got to be very lucky to walk in and just get the right remedy." So the power is, and is not, in the pill: "P, and not-P," as philosophers of logic would say.

If they can't finesse it with the "power is not in the pill" para-dox, how else do the homeopaths get around all this negative data? Dr. Thompson—from what I have seen—is a fairly clear-thinking and civilized homeopath. She is, in many respects, alone. Homeo-paths have been careful to keep themselves outside the civilizing environment of the university, where the influence and question-ing of colleagues can help refine ideas and weed out the bad ones. In their rare forays, they enter them secretively, walling themselves and their ideas off from criticism or review, refusing to share even what is in their exam papers with outsiders.

It is rare to find a homeopath engaging on the issue of the evi-dence, but what happens when they do? I can tell you. They get an-gry; they threaten to sue; they scream and shout at you at meetings; they complain spuriously and with ludicrous misrepresentations—time-consuming to expose, of course, but that's the point of harassment—to the Press Complaints Commission and your editor; they send hate mail and accuse you repeatedly of somehow being in the pocket of big pharma (falsely, although you start to wonder why you bother having principles when faced with this kind of be-havior). They bully, they smear, to the absolute top of the profes-sion, and they do anything they can in a desperate bid to *shut you up* and avoid having a discussion about the evidence. They have even been known to threaten violence (I won't go into it here, but I manage these issues extremely seriously).

I'm not saying I don't enjoy a bit of banter. I'm just pointing out that you don't get anything quite like this in most other fields, and homeopaths, among all the people in this book, with the ex-ception of the odd nutritionist, seem to me to be a uniquely angry breed. Experiment for yourself by chatting with them about evi-dence, and let me know what you find.

By now your head is hurting, because of all those mischievous, confusing homeopaths and their weird, labyrinthine defenses; you need a lovely science massage. Why is evidence so complicated?

Why do we need all these clever tricks, these special research paradigms? The answer is easy: the world is much more complicated than simple stories about pills making people get better. We are human, we are irrational, we have foibles, and the power of the mind over the body is greater than anything you have previously imagined.

5

THE PLACEBO EFFECT

For all the dangers of complementary and alternative medicine, to me the greatest disappointment is the way it distorts our understanding of our bodies. Just as the big bang theory is far more interesting than the creation story in Genesis, so the story that science can tell us about the natural world is far more interesting than any fable about magic pills concocted by an alternative therapist. To redress that balance, I'm offering you a whirlwind tour of one of the most bizarre and enlightening areas of medical research: the relationship between our bodies and our minds, the role of meaning in healing, and in particular the placebo effect.

Much like quackery, placebos became unfashionable in medicine once the biomedical model started to produce tangible results. An editorial in 1890 sounded its death knell, describing the case of a doctor who had injected his patient with water instead of morphine; she recovered perfectly well, but then discovered the deception, disputed the bill in court, and won. The editorial was a lament, because doctors have known that reassurance and a good bedside manner can be very effective for as long as medicine has existed. "Shall [the placebo] never again have an opportunity of exerting its wonderful psychological effects as faithfully as one of its more toxic conveners?" asked the *Medical Press* at the time.

Luckily, its use survived. Throughout history, the placebo effect has been particularly well documented in the field of pain, and some of the stories are striking. Henry Beecher, an American anesthetist, wrote about operating on a soldier with horrific injuries in a World War II field hospital, using salt water because the morphine was all gone, and to his astonishment the patient was fine. Peter Parker, an American missionary, described performing surgery without anesthesia on a Chinese patient in the mid-nineteenth century; after the operation, she "jumped upon the floor," bowed, and walked out of the room as if nothing had happened.

Theodor Kocher performed sixteen hundred thyroidectomies without anesthesia in Switzerland in the 1890s, and I take my hat off to a man who can do complicated neck operations on conscious patients. Mitchel in the early twentieth century was performing full amputations and mastectomies, entirely without anesthesia; and surgeons from before the invention of anesthesia often described how some patients could tolerate knife cutting through muscle, and saw cutting through bone, perfectly awake and without even clenching their teeth. You might be tougher than you think.

These are just stories, and the plural of "anecdote" is not data. Everyone knows about the power of the mind—whether it's stories of mothers enduring biblical pain to avoid dropping a boiling kettle on their babies or people lifting cars off their girlfriends like the Incredible Hulk—but devising an experiment that teases the psychological and cultural benefits of a treatment away from the biomedical effects is trickier than you might think. After all, what do you compare a placebo against? Another placebo? Or no treatment at all?

THE PLACEBO ON TRIAL

In most studies we don't have a "no treatment" group to compare both the placebo and the drug with, and for a very good ethical reason: if your patients are ill, you shouldn't be leaving them untreated simply because of your own mawkish interest in the placebo effect. In fact, in most cases today it is considered wrong even to use a placebo in a trial; whenever possible you should compare your new treatment with the best preexisting, current treatment.

This is not just for ethical reasons (although it is enshrined in the Declaration of Helsinki, the international ethics bible). Placebo-controlled trials are also frowned upon by the evidence-based medicine community, because it knows it's an easy way to cook the books and get easy positive trial data to support your company's big new investment. In the real world of clinical practice, patients and doctors aren't so interested in whether a new drug works better than *nothing*; they're interested in whether it works *better than the best treatment they already have*.

There have been occasions in medical history when researchers were more cavalier. The Tuskegee Syphilis Study, for example, is one of America's most shaming hours: 399 poor, rural African-American men were recruited by the U.S. Public Health Service in 1932 for an observational study to see what happened if syphilis was left, very simply, untreated. Astonishingly, the study ran right through to 1972. In 1949 penicillin was introduced as an effective treatment for syphilis. These men did not receive that drug, nor did they receive Salvarsan, nor indeed did they receive an apology until 1997, from Bill Clinton.

If we don't want to do unethical scientific experiments with "no treatment" groups on sick people, how else can we determine the size of the placebo effect on modern illnesses? First, and rather ingeniously, we can compare one placebo with another.

The first experiment in this field was a meta-analysis by Daniel Moerman, an anthropologist who has specialized in the placebo effect. He took the trial data from placebo-controlled trials of gastric ulcer medication, which was his first cunning move, because gastric ulcers are an excellent thing to study: their presence or absence is determined very objectively, with a gastroscopy camera passed down into the stomach, to avoid any doubt.

Moerman took only the placebo data from these trials, and then, in his second ingenious move, from all these studies, of all the different drugs, with their different dosing regimes, he took the ulcer-healing rate from the placebo arm of trials in which the placebo treatment was two sugar pills a day, and compared that with the ulcer-healing rate in the placebo arm of trials in which the placebo was four sugar pills a day. He found, spectacularly, that four sugar pills are better than two (these findings have also been replicated in a different data set, for those who are switched on enough to worry about the replicability of important clinical findings).

WHAT THE TREATMENT LOOKS LIKE

So four pills are better than two, but how can this be? Does a placebo sugar pill simply exert an effect like any other pill? Is there a dose response curve, as pharmacologists would find for any other drug? The answer is that the placebo effect is about far more than just the pill; it is about the cultural meaning of the treatment. Pills don't simply manifest themselves in your stomach; they are given in particular ways, they take varying forms, and they are swallowed with expectations, all of which have an impact on a person's beliefs about his own health and, in turn, on outcome. Homeopathy is, for example, a perfect example of the value in ceremony.

I understand this might well seem improbable to you, so I've corralled some of the best data on the placebo effect into one place, and the challenge is this: see if you can come up with a better explanation for what is, I guarantee, a seriously strange set of experimental results.

First up, Blackwell (1972) did a set of experiments on fifty-seven college students to determine the effect of color—as well as the number of tablets—on the effects elicited. The subjects were sitting through a boring hourlong lecture and were given either one or two pills, which were either pink or blue. They were told that they could expect to receive either a stimulant or a sedative. Since these were psychologists, and this was back when you could do whatever you wanted to your subjects—even lie to them—the treatment that *all* the students received consisted simply of sugar pills, but of different colors.

Afterward, when they measured alertness—as well as any subjective effects—the researchers found that two pills were more effective than one, as we might have expected (and two pills were better at eliciting side effects too). They also found that color had an effect on outcome: the pink sugar tablets were better at maintaining concentration than the blue ones. Since colors in themselves have no intrinsic pharmacological properties, the difference in effect could only be due to the cultural meanings of pink and blue: pink is alerting; blue is cool. Another study suggested that oxazepam, a drug similar to Valium (which was once unsuccessfully prescribed by our doctor for me as a hyperactive child) was more effective at treating anxiety in a green tablet and more effective for depression when yellow.

Drug companies, more than most, know the benefits of good branding; they spend more on PR, after all, than they do on research and development. As you'd expect from men of action with large houses in the country, they put these theoretical ideas into practice, so Prozac, for example, is white and blue, and in case you think I'm cherry-picking here, a survey of the color of pills currently on the market found that stimulant medication tends to come in red, orange, or yellow tablets, while antidepressants and tranquilizers are generally blue, green, or purple.

Issues of form go much deeper than color. In 1970 a sedative—chlordiazepoxide—was found to be more effective in capsule form than pill form, even for the very same drug, in the very same dose; capsules at the time felt newer, somehow, and more sciencey. Maybe you've caught yourself splashing out and paying extra for ibuprofen capsules in the pharmacy.

Route of administration has an effect as well: saltwater injections have been shown in three separate experiments to be more effective than sugar pills for blood pressure, for headaches, and for postoperative pain, not because of any physical benefit of saltwater injection over sugar pills—there isn't one—but because, as everyone knows, an injection is a much more dramatic intervention than just taking a pill.

Closer to home for the alternative therapists, the *British Medical Journal* recently published an article comparing two different placebo treatments for arm pain, one of which was a sugar pill, and one of which was a ritual, a treatment modeled on acupuncture. The trial found that the more elaborate placebo ritual had a greater benefit.

But the ultimate testament to the social construction of the placebo effect must be the bizarre story of packaging. Pain is an area where you might suspect that expectation would have a particularly significant effect. Most people have found that they can take their minds off pain—to at least some extent—with distraction, or have had a toothache that got worse with stress.

Branthwaite and Cooper did a truly extraordinary study in 1981, looking at 835 women with headaches. It was a four-armed study, in which the subjects were given either aspirin or placebo pills, and these pills in turn were packaged either in blank, bland, neutral boxes or in full, flashy, brand-name packaging. They found—as you'd expect—that aspirin had more of an effect on headaches than sugar pills, but more than that, they found that the packaging itself had a beneficial effect, enhancing the benefit of both the placebo and the aspirin.

People I know still insist on buying brand-name painkillers. As you can imagine, I've spent half my life trying to explain to them why this is a waste of money, but in fact, the paradox of Branthwaite and Cooper's experimental data is that they were right all along. Whatever pharmacology theory tells you, that brand-named version *is* better, and there's just no getting away from it. Part of that might be the cost; a recent study looking at pain caused by electric shocks showed that a pain relief treatment was stronger when subjects were told it cost $2.50 than when they were told it cost 10 cents. (And a paper currently in press shows that people are more likely to take advice when they have paid for it.)

It gets better—or worse, depending on how you feel about your worldview slipping sideways. Montgomery and Kirsch (1996) told college students they were taking part in a study on a new local anesthetic called trivaricaine. Trivaricaine is brown, you paint it on your skin, it smells like a medicine, and it's so potent you have to wear gloves when you handle it: or that's what they implied to the students. In fact, it's made of water, iodine, and thyme oil (for the smell), and the experimenters (who also wore white coats) were using rubber gloves only for a sense of theater. None of these ingredients will affect pain.

The trivaricaine was painted onto one or other of the subjects' index fingers, and the experimenters then applied painful pressure with a vise. One after another, in varying orders, pain was applied, trivaricaine was applied, and as you would expect by now, the subjects reported less pain, and less unpleasantness, for the fingers that were pretreated with the amazing trivaricaine. This is a placebo effect, but the pills have gone now.

It gets stranger. Sham ultrasound is beneficial for dental pain, placebo operations have been shown to be beneficial in knee pain (the surgeon just makes fake keyhole surgery holes in the side and mucks about for a bit as if she were doing something useful), and placebo operations have even been shown to improve angina.

That's a pretty big deal. Angina is the pain you get when there's not enough oxygen getting to your heart muscle for the work it's doing. That's why it gets worse with exercise: because you're demanding more work from the heart muscle. You might get a similar pain in your thighs after bounding up ten flights of stairs, depending on how fit you are.

Treatments that help angina usually work by dilating the blood vessels to the heart, and a group of chemicals called nitrates are used for this purpose very frequently. They relax the smooth muscle in the body, dilating the arteries so more blood can get through (they also relax other bits of smooth muscle in the body, including your anal sphincter, which is why a variant is sold as "liquid gold" in sex shops).

In the 1950s there was an idea that you could get blood vessels in the heart to grow back, and thicker, if you tied off an artery on the front of the chest wall that wasn't very important, but that branched off the main heart arteries. The idea was that this would send messages back to the main branch of the artery, telling it that more artery growth was needed, so the body would be tricked.

Unfortunately this idea turned out to be nonsense, but only after a fashion. In 1959 a placebo-controlled trial of the operation was performed: in some operations they did the whole thing properly, but in the "placebo" operations they went through the motions but didn't tie off any arteries. It was found that the placebo operation was just as good as the real one—people seemed to get a bit better in both cases, and there was little difference between the groups—but the strangest thing about the whole affair was that nobody made a fuss at the time. The real operation wasn't any better than a sham operation, sure, but how could we explain the

fact that people had been sensing an improvement from the operation for a very long time? Nobody thought of the power of placebo.
The operation was simply binned.

That's not the only time a placebo benefit has been found at
the more dramatic end of the medical spectrum. A Swedish study
in the late 1990s showed that patients who had pacemakers in-
stalled but not switched on did better than they had been doing
before (although they didn't do as well as people with working pacemakers inside them, to be clear). Even more recently, one study of a
very hi-tech "angioplasty" treatment, involving a large and sciencey-
looking laser catheter, showed that sham treatment was almost as
effective as the full procedure.

"Electrical machines have great appeal to patients," wrote Dr.
Alan Johnson in *The Lancet* in 1994 about this trial, "and recently
anything to do with the word LASER attached to it has caught
the imagination." He's not wrong. I went to visit an alternative
therapist once, and she did gem therapy on me, with a big shiny
science machine that shone different-colored beams of light onto
my chest. It's hard not to see the appeal of things like gem therapy
in the context of the laser catheter experiment. In fact, the way
the evidence is stacking up, it's hard not to see all the claims of
alternative therapists, for all their wild, wonderful, authoritative,
and empathic interventions, in the context of this chapter.

In fact, even the lifestyle gurus get a look in, in the form of an
elegant study that examined the effect of simply being told that
you are doing something healthy. Eighty-four female room atten-
dants working in various hotels were divided into two groups.
One group was told that cleaning hotel rooms is "good exercise"
and "satisfies the Surgeon General's recommendations for an active lifestyle," along with elaborate explanations of how and why;
the "control" group did not receive this cheering information
and just carried on cleaning hotel rooms. Four weeks later, the
"informed" group perceived themselves to be getting significantly

more exercise than before and showed a significant decrease in weight, body fat, waist-to-hip ratio, and body mass index, but amazingly, both groups were still reporting the same amount of activity.*

WHAT THE DOCTOR SAYS

If you can believe fervently in your treatment, even though controlled tests show that it is quite useless, then your results are much better, your patients are much better, and your income is much better too. I believe this accounts for the remarkable success of some of the less gifted, but more credulous members of our profession, and also for the violent dislike of statistics and controlled tests which fashionable and successful doctors are accustomed to display.

—Richard Asher, *Talking Sense*, Pitman Medical, 1972

As you will now be realizing, in the study of expectation and belief, we can move away from pills and devices entirely. It turns out, for example, that both what the doctor says and what the doctor believes have an effect on healing. If that sounds obvious, I should say they have an effect that has been measured, elegantly, in carefully designed trials.

Gryll and Katahn (1978) gave patients a sugar pill before a dental injection, but the doctors who were handing out the pill gave it in one of two different ways: either with an outrageous oversell ("This is a recently developed pill that's been shown to be very effective . . . effective almost immediately . . .") or downplayed, with an undersell ("This is a recently developed pill . . . personally I've

---

*I agree: this is a bizarre and outrageous experimental finding, and if you have a good explanation for how it might have come about, the world would like to hear from you. Follow the reference, read the full paper online, and start a blog, or write a letter to the journal that published it.

not found it to be very effective . . ."). The pills that were handed out with the positive message were associated with less fear, less anxiety, and less pain.

Even if he says nothing, what the doctor knows can affect treatment outcomes; the information leaks out, in mannerisms, affect, eyebrows, and nervous smiles, as Gracely (1985) demonstrated with a truly ingenious experiment, although understanding it requires a tiny bit of concentration.

He took patients having their wisdom teeth removed, and split them randomly into three treatment groups: they would have salt water (a placebo that does "nothing," at least not physiologically) or fentanyl (an excellent opiate painkiller, with a black-market retail value to prove it), or naloxone (an opiate receptor blocker that would actually increase the pain).

In all cases the doctors were blinded to which of the three treatments they were giving to each patient, but Gracely was *really* studying the effect of his doctors' beliefs, so the groups were further divided in half again. In the first group, the doctors giving the treatment were told, truthfully, that they could be administering placebo, or naloxone, or the pain-relieving fentanyl; this group of doctors knew there was a chance that they were giving something that would reduce pain.

In the second group, the doctors were lied to; they were told they were giving either placebo or naloxone, two things that could only do nothing or actively make the pain worse. But in fact, without the doctors' knowledge, some of their patients were actually getting the pain-relieving fentanyl. As you would expect by now, just through manipulation of what the *doctors believed* about the injections they were giving, even though they were forbidden from vocalizing their beliefs to the patients, there was a difference in outcome between the two groups. The first group experienced significantly less pain. This difference had nothing to do with what actual medicine was being given or even with what information the patients knew; it was entirely down to what the doctors knew. Perhaps they winced when they gave the injection. I think you might have.

PLACEBO EXPLANATIONS

Even if they do nothing, doctors, by their manner alone, can reassure. And even reassurance can in some senses be broken down into informative constituent parts. In 1987, Thomas showed that simply giving a diagnosis—even a fake "placebo" diagnosis—improved patient outcomes. Two hundred patients with abnormal symptoms, but no signs of any concrete medical diagnoses, were divided randomly into two groups. The patients in one group were told, "I cannot be certain of what the matter is with you," and two weeks later only 39 percent were better; the other group was given a firm diagnosis, with no messing about, and confidently told they would be better within a few days. Sixty-four percent of that group got better in two weeks.

This raises the specter of something way beyond the placebo effect, and cuts even further into the work of alternative therapists, because we should remember that alternative therapists don't just give placebo treatments; they also give what we might call placebo explanations or placebo diagnoses: ungrounded, unevidenced, often fantastical assertions about the nature of the patient's disease, involving magical properties, or energy, or supposed vitamin deficiencies, or "imbalances," which the therapist claims uniquely to understand.

And here it seems that this placebo explanation—even if grounded in sheer fantasy—can be beneficial to a patient, although interestingly, perhaps not without collateral damage, and it must be done delicately; assertively and authoritatively giving someone access to the sick role can also reinforce destructive illness beliefs and behaviors, unnecessarily medicalize symptoms like aching muscles (which for many people are everyday occurrences), and militate against people's getting on with life and getting better. It's a very tricky area.

I could go on. In fact, there has been a huge amount of research into the value of a good therapeutic relationship, and the general finding is that doctors who adopt a warm, friendly, and reassuring manner are more effective than those who keep consultations formal and do not offer reassurance. In the real world, there are structural cultural changes that make it harder and harder for a medical doctor to maximize the therapeutic benefit of a consultation. First, there is the pressure on time; a doctor can't do much in a six-minute appointment.

But more than these practical restrictions, there have also been structural changes in the ethical presumptions made by the medical profession, which make reassurance an increasingly outré business. A modern medic would struggle to find a form of words that would permit her to hand out a placebo, for example, and this is because of the difficulty in resolving two very different ethical principles: one is our obligation to heal our patients as effectively as we can; the other is our obligation not to tell them lies. In many cases the prohibition on reassurance and smoothing over worrying facts has been formalized, as the doctor and philosopher Raymond Tallis recently wrote, beyond what might be considered proportionate: "The drive to keep patients fully informed has led to exponential increases in the formal requirements for consent that only serve to confuse and frighten patients while delaying their access to needed medical attention."

I don't want to suggest for one moment that historically this was the wrong call. Surveys show that patients want their doctors to tell them the truth about diagnoses and treatments.

What is odd, perhaps, is how the primacy of patient autonomy and informed consent over efficacy, which is what we're talking about here, was presumed but not actively discussed within the medical profession. Although the authoritative and paternalistic reassurance of the Victorian doctor who "blinds with science" is a thing of the past in medicine, the success of the alternative therapy movement—practitioners mislead, mystify, and blind their patients with sciencey-sounding "authoritative" explanations, like the most patronizing Victorian doctor imaginable—suggests that there may still be a market for that kind of approach.

About a hundred years ago, these ethical issues were carefully documented by a thoughtful native Canadian Indian called Quesalid. Quesalid was a skeptic. He thought shamanism was bunk, that it worked only through belief, and he went undercover to investigate this idea. He found a shaman who was willing to take him on, and he learned all the tricks of the trade, including the classic performance piece in which the healer hides a tuft of down in the corner of his mouth and then, sucking and heaving, right at the peak of his healing ritual, brings it up, covered in blood from where he has discreetly bitten his lip, and solemnly presents it to the onlookers as a pathological specimen, extracted from the body of the afflicted patient.

Quesalid had proof of the fakery, he knew the trick as an insider and was all set to expose those who carried it out; but as part of his training he had to do a bit of clinical work, and he was summoned by a family "who had dreamed of him as their saviour" to see a patient in distress. He did the trick with the tuft and was appalled, humbled, and amazed to find that his patient got better.

Although he continued to maintain a healthy skepticism about most of his colleagues, Quesalid, to his own surprise, perhaps, went on to have a long and productive career as a healer and shaman. The anthropologist Claude Lévi-Strauss, in his paper "The Sorcerer and His Magic," doesn't quite know what to make of it, "but it is evident that Quesalid carries on his craft conscientiously, takes pride in his achievements, and warmly defends the technique of the bloody down against all rival schools. He seems to have completely lost sight of the fallaciousness of the technique that he had so disparaged at the beginning."

Of course, it may not even be necessary to deceive your patient in order to maximize the placebo effect; a classic study from 1965—albeit small and without a control group—gives a small hint of what might be possible here. The researchers gave a pink placebo sugar pill three times a day to "neurotic" patients, with good effect, and the explanation given to the patients was startlingly clear about what was going on:

A script was prepared and carefully enacted as follows: "Mr. Doe . . . we have a week between now and your next appointment, and we would like to do something to give you some relief from your symptoms. Many different kinds of tranquilizers and similar pills have been used for conditions such as yours, and many of them have helped. Many people with your kind of condition have also been helped by what are sometimes called 'sugar pills,' and we feel that a so-called sugar pill may help you, too. Do you know what a sugar pill is? A sugar pill is a pill with no medicine in it at all. I think this pill will help you as it has helped so many others. Are you willing to try this pill?"

The patient was then given a supply of placebo in the form of pink capsules contained in a small bottle with a label showing the name of the Johns Hopkins Hospital. He was instructed to take the capsules quite regularly, one capsule three times a day at each meal time.

The patients improved considerably. I could go on, but this all sounds a bit wishy-washy. We all know that pain has a strong psychological component. What about the more robust stuff, something more counterintuitive, something more . . . sciencey?

Dr. Stewart Wolf took the placebo effect to the limit. He took two women who were suffering with nausea and vomiting, one of them pregnant, and told them he had a treatment that would improve their symptoms. In fact, he passed a tube down into their stomachs (so that they wouldn't taste the revolting bitterness) and administered ipecac, a drug that should actually *induce* nausea and vomiting.

Not only did the patients' symptoms improve, but their gastric contractions, which ipecac should worsen, were *reduced*. His results suggest—albeit it in a very small sample—that a drug could be made to have the opposite effect from what you would predict from the pharmacology, simply by manipulating people's expectations. In this case, the placebo effect outgunned even the pharmacological influences.

MORE THAN MOLECULES?

So is there any research from the basic science of the laboratory bench to explain what's happening when we take a placebo? Well, here and there, yes, although they're not easy experiments to do. It's been shown, for example, that the effects of a real drug in the body can sometimes be induced by the placebo "version," not only in humans but also in animals. Most drugs for Parkinson's disease work by increasing dopamine release; patients receiving a placebo treatment for Parkinson's disease, for example, showed extra dopamine release in the brain.

Zubieta (2005) showed that subjects who are subjected to pain and then given a placebo release more endorphins than people who got nothing. (I feel duty bound to mention that I'm a bit dubious about this study, because the people on placebo also endured more painful stimuli, another reason why they might have had higher endorphins; consider this a small window into the wonderful world of interpreting uncertain data.)

If we delve further into theoretical work from the animal kingdom, we find that animals' immune systems can be conditioned to respond to placebos, in exactly the same way that Pavlov's dog began to salivate in response to the sound of a bell. Researchers have measured immune system changes in dogs using just flavored sugar water, once that flavored water has been associated with immunosuppression, by administering it repeatedly alongside cyclophosphamide, a drug that suppresses the immune system.

A similar effect has been demonstrated in humans when the researchers gave healthy subjects a distinctively flavored drink at the same time as cyclosporine A (a drug that measurably reduces your immune function). Once the association was set up with sufficient repetition, they found that the flavored drink on its own could induce modest immune suppression. Researchers have even managed to elicit an association between sherbet and natural killer cell activity.

What does this all mean for you and me?

People have tended to think, rather pejoratively, that if your pain responds to a placebo, that means it's "all in the mind." From survey data, even doctors and nurses buy into this canard. An article from *The Lancet* in 1954—another planet in terms of how doctors spoke about patients—states that "for some unintelligent or inadequate patients, life is made easier by a bottle of medicine to comfort the ego."

This is wrong. It's no good trying to exempt yourself, and pretend that this is about other people, because we all respond to the placebo. Researchers have tried hard in experiments and surveys to characterize placebo responders, but the results overall come out like a horoscope that could apply to everybody: placebo responders have been found to be more extroverted but more neurotic, more well adjusted but more antagonistic, more socially skilled, more belligerent but more acquiescent, and so on. The placebo responder is everyman. You are a placebo responder. Your body plays tricks on your mind. You cannot be trusted.

How do we draw all this together? Moerman reframes the placebo effect as the meaning response—"the psychological and physiological effects of meaning in the treatment of illness"—and it's a compelling model. He has also performed one of the most impressive quantitative analyzes of the placebo effect and how it changes with context, again on stomach ulcers. As we've said before, this is an excellent disease to study, because ulcers are prevalent and treatable, but most important because treatment success can be unambiguously recorded by having a look down there with a gastroscope.

Moerman examined 117 studies of ulcer drugs from between 1975 and 1994 and found, astonishingly, that they interact in a way you would never have expected: culturally, rather than pharmacodynamically. Cimetidine was one of the first ulcer drugs on the market, and it is still in use today; in 1975, when it was new, it eradicated 80 percent of ulcers, on average, in the various different trials. As time passed, however, the success rate of cimetidine deteriorated to just 50 percent. Most interestingly, this deterioration seems to have occurred particularly after the introduction of ranitidine, a competing and supposedly superior drug, onto the market five years later. So the selfsame drug became less effective with time, as new drugs were brought in.

There are a lot of possible interpretations of this. It's possible, of course, that it was a function of changing research protocols. But a highly compelling possibility is that the older drugs became less effective after new ones were brought in because of deteriorating medical belief in them. Another study from 2002 looked at seventy-five trials of antidepressants over the previous twenty years and found that the response to placebo had increased significantly in recent years (as had the response to medication), perhaps as our expectations of those drugs increased.

Findings like these have important ramifications for our view of the placebo effect, and for all medicine, since it may be a potent universal force. We must remember, specifically, that the placebo effect—or the meaning effect—is *culturally specific*. Brand-name painkillers might be better than blank-box painkillers over here, but if you went and found someone with toothache in 6000 B.C., or up the Amazon in 1880, or dropped in on Soviet Russia during the 1970s, where nobody had seen the TV advert with the attractive woman wincing from a pulsing red orb of pain in her forehead, who swallows the painkiller, and then the smooth, reassuring blue suffuses her body . . . In a world without those cultural preconditions to set up the dominoes, you would expect aspirin to do the same job no matter what box it came out of.

This also has interesting implications for the transferability of alternative therapies. The British novelist Jeanette Winterson, for example, has written in *The Times* (London) trying to raise money for a project to treat AIDS sufferers in Botswana—where a quarter of the population is HIV positive—with homeopathy. We must put aside the irony here of taking homeopathy to a country that has been engaged in a water war with neighboring Namibia, and we must also let lie the tragedy of Botswana's devastation by AIDS, which is so phenomenal—I'll say it again: *a quarter of the population is HIV positive*—that if it is not addressed rapidly and robustly, the entire economically active portion of the population could simply cease to exist, leaving what would be effectively a noncountry.

All this tragedy left aside, what's interesting for our purposes is the idea that you could take your Western, individualistic, patient-empowering, antimedical establishment, and very culturally specific placebo to a country with so little health care infrastructure and expect it to work all the same. The greatest irony of all is that if homeopathy has any benefits at all for AIDS sufferers in Botswana, it may be through its implicit association with the white-coat Western medicine that so many African countries desperately need.

So, if you go off now and chat to an alternative therapist about the contents of this chapter—as I very much hope you will—what will you hear? Will he smile, nod, and agree that his rituals have been carefully and elaborately constructed over many centuries of trial and error to elicit the best placebo response possible? That there are more fascinating mysteries in the true story of the relationship between body and mind than any fanciful notion of quantum energy patterns in a sugar pill?

To me, this is yet another example of a fascinating paradox in the philosophy of alternative therapists: when they claim that their treatments are having a specific and measurable effect on the body, through specific technical mechanisms rather than ritual, they are championing a very old-fashioned and naive form of biological reductionism, in which the mechanics of their interventions, rather than the relationship and the ceremony, have the positive effect on healing. Once again, it's not just that they have no evidence for their claims about how their treatments work: it's that their claims are mechanistic, intellectually disappointing, and simply less interesting than the reality.

AN ETHICAL PLACEBO?

But more than anything, the placebo effect throws up fascinating ethical quandaries and conflicts around our feelings on pseudoscience. Let's take our most concrete example so far: Are the sugar pills of homeopathy exploitative if they work only as a placebo? A pragmatic clinician could only consider the value of a treatment by considering it in context.

Here is a clear example of the benefits of placebo. During the nineteenth-century cholera epidemic in London, deaths were occurring in the London Homeopathic Hospital at just one-third of the rate as in the Middlesex Hospital, but a placebo effect is unlikely to be all that beneficial in this condition. The reason for homeopathy's success in this case is more interesting: at the time, nobody could treat cholera. So while hideous medical practices such as bloodletting were actively harmful, the homeopaths' treatments at least did nothing either way.

Today, similarly, there are often situations where people want treatment, but medicine has little to offer—lots of back pain, stress at work, medically unexplained fatigue, and most common colds, to give just a few examples. Going through a theater of medical treatment, and trying every medication in the book, will give you only side effects. A sugar pill in these circumstances seems a very sensible option, as long as it can be administered cautiously, and ideally with a minimum of deceit.

But just as homeopathy has unexpected benefits, so it can have unexpected side effects. Believing in things that have no evidence carries its own corrosive intellectual side effects, just as prescribing a pill in itself carries risks: it medicalizes problems, as we shall see, it can reinforce destructive beliefs about illness, and it can promote the idea that a pill is an appropriate response to a social problem, or a modest viral illness.

There are also more concrete harms, specific to the culture in which the placebo is given, rather than the sugar pill itself. For example, it's routine marketing practice for homeopaths to denigrate mainstream medicine. There's a simple commercial reason for this: survey data shows that a disappointing experience with mainstream medicine is almost the only factor that regularly correlates with choosing alternative therapies. This is not just talking medicine down; one study found that more than half of all the homeopaths in the U.K. approached advised patients against the MMR vaccine for their children, acting irresponsibly on what will quite probably come to be known as the media's MMR hoax. How did the alternative therapy world deal with this concerning finding, that so many among them were quietly undermining the vaccination schedule? Prince Charles's office tried to have the lead researcher into the matter sacked.

Angry People

A BBC *Newsnight* investigation found that almost all the homeopaths approached recommended ineffective homeopathic pills to protect against malaria, and advised against medical malaria prophylactics, while not even giving basic advice on mosquito bite prevention. This may strike you as neither holistic nor "complementary." How did the self-proclaimed "regulatory bodies" in homeopathy deal with this? None took any action against the homeopaths concerned.

And at the extreme, when they're not undermining public health campaigns and leaving their patients exposed to fatal diseases, homeopaths who are not medically qualified can miss fatal diagnoses or actively disregard them, telling their patients grandly to stop using their inhalers and to throw away their heart pills. There are plenty of examples, but I have too much style to document them here. Suffice to say that while there may be a role for an ethical placebo, homeopaths, at least, have ably demonstrated that they have neither the maturity nor the professionalism to provide it. Fashionable doctors, meanwhile, stunned by the commercial appeal of sugar pills, sometimes wonder—rather unimaginatively—whether they should simply get in on the act and sell some themselves. A smarter idea by far, surely, is to exploit the research we have seen, but only to enhance treatments that really *do* perform better than placebo and improve health care without misleading our patients.

6

THE NONSENSE DU JOUR

Now we need to raise our game. Food has become an international obsession. The newspapers sometimes seem to be engaged in a bizarre ongoing ontological project, diligently sifting through all the inanimate objects of the universe in order to categorize them as a cause of—or cure for—cancer. At the core of this whole project are a small number of repeated canards, basic misunderstandings of evidence that recur with phenomenal frequency. These intellectual crimes are ferried to you by journalists, celebrities, and, of course, "nutritionists," members of a newly invented profession who must create a commercial space to justify their own existence. In order to do this, they must mystify and overcomplicate diet and foster your dependence upon them. Their profession is based on a set of very simple mistakes in how we interpret scientific literature: they extrapolate wildly from "laboratory bench data" to make claims about humans; they extrapolate from "observational data" to make "intervention claims"; they "cherry-pick"; and last, they quote published scientific research evidence that seems, as far as one can tell, not to exist.

It's worth going through these misrepresentations of evidence, mainly because they are fascinating illustrations of how people can get things wrong, but also because the aim of this book is that you should be future-proofed against new variants of bullshit. There are also two things we should be very clear on. First, I'm picking out individual examples as props, but these are characteristic of the genre; I could have used many more. Nobody is being bullied, and none of them should be imagined to stand out from the nutrition- ist crowd, although I'm sure some of the people covered here won't be able to understand how they've done anything wrong.

Second, I am not deriding simple, sensible, healthy eating ad- vice. A straightforwardly healthy diet, along with many other as- pects of lifestyle (many of which are probably more important, not that you'd know it from reading the papers), is very important. But the media nutritionists speak beyond the evidence. Often it is about selling pills; sometimes it is about selling dietary fads, or new diag- noses, or fostering dependence, but it is always driven by their de- sire to create a market for themselves, in which they are the expert, whereas you are merely bamboozled and ignorant.

Prepare to switch roles.

THE FOUR KEY ERRORS

DOES THE DATA EXIST?

This is perhaps the simplest canard of all, and it happens with surprising frequency, in some rather authoritative venues. Here is Michael van Straten on BBC *Newsnight*, talking "fact." If you pre- fer not to take it on faith that his delivery is earnest, definitive, and perhaps even slightly patrician, you can watch the clip online, and in reality, you can see this kind of thing on any station, with tedious frequency.

"When Michael van Straten started writing about the magical medicinal powers of fruit juices, he was considered a crank," *News- night* begins. "But now he finds he's at the forefront of fashion." Van

Straten hands the reporter a glass of juice. "Two years added to your life expectancy in that!" He chuckles; then a moment of seriousness: "Well, six months, being honest about it." A correction. "A recent study just published last week in America showed that eating pomegranates, pomegranate juice, can actually protect you against aging, against wrinkles," he says.

Hearing this, the viewer might naturally conclude that a study has recently been published in America showing that pomegranates can protect against aging. But if you go to Medline, the standard search tool for finding medical academic papers, no such study exists, or at least not that I can find. Perhaps there's some kind of leaflet from the pomegranate industry doing the rounds. Van Straten goes on: "There's a whole group of plastic surgeons in the States who've done a study giving some women pomegranates to eat, and juice to drink, after plastic surgery and before plastic surgery: and they heal in half the time, with half the complications, and no visible wrinkles!" Again, it's a very specific claim—a human trial on pomegranates and surgery—and again, there is nothing in the studies database.

So could you fairly characterize this *Newsnight* performance as "lying"? Absolutely not. In defense of almost all nutritionists, I would argue that they lack the academic experience, the ill will, and perhaps even the intellectual horsepower necessary to be fairly derided as liars. The philosopher professor Harry Frankfurt of Princeton University discusses this issue at length in his classic 1986 essay "On Bullshit." Under his model, "bullshit" is a form of falsehood distinct from lying: the liar knows and cares about the truth but deliberately sets out to mislead; the truth speaker knows the truth and is trying to give it to us; the bullshitter, meanwhile, does not care about the truth and is simply trying to impress us:

> It is impossible for someone to lie unless he thinks he knows the truth. Producing bullshit requires no such conviction . . . When an honest man speaks, he says only

what he believes to be true; and for the liar, it is correspondingly indispensable that he considers his statements to be false. For the bullshitter, however, all these bets are off: he is neither on the side of the true nor on the side of the false. His eye is not on the facts at all, as the eyes of the honest man and of the liar are, except insofar as they may be pertinent to his interest in getting away with what he says. He does not care whether the things he says describe reality correctly. He just picks them out, or makes them up, to suit his purpose.

I see van Straten, like many of the subjects in this book, very much in the "bullshitting" camp. Is it unfair for me to pick out this one man? Perhaps. In biology fieldwork, you throw a wired square called a quadrat at random out onto the ground, and then examine whatever species fall underneath it. This is the approach I have taken with nutritionists, and until I have a Department of Pseudoscience Studies with an army of Ph.D. students doing quantitative work on who is the worst, we shall never know. Van Straten seems like a nice, friendly guy. But we have to start somewhere.

OBSERVATION OR INTERVENTION?

Does the rooster's crow cause the sun to rise? No. Does this light switch make the room get brighter? Yes. Things can happen at roughly the same time, but that is weak, circumstantial evidence for causation. Yet it's exactly this kind of evidence that is used by media nutritionists as confident proof of their claims in our second major canard.

Here is an interesting example from Angela Dowden, who, according to *The Daily Mirror*, is "Britain's leading nutritionist." In one of her columns in the *Mirror* explaining which foods offer protection from the sun during a heat wave, she writes: "An Australian study in 2001 found that olive oil (in combination with fruit, vegetables and pulses) offered measurable protection against skin wrinkling. Eat more olive oil by using it in salad dressings or dip bread in it rather than using butter."

That's very specific advice, with a very specific claim, quoting a very specific reference, and with a very authoritative tone. It's typical of what you get in the papers from media nutritionists. Let's go to the library and fetch out the paper she refers to ("Skin wrinkling: can food make a difference?" Purba, M. B., et al. *Journal of American College of Nutrition* 20, no. 1 [February, 2001]: 71–80). Before we go any further, we should be clear that we are criticizing Dowden's *interpretation* of this research, not the research itself, which we assume is a faithful description of the investigative work that was done.

This was an observational study, not an intervention study. It did not give people olive oil for a time and then measure differences in wrinkles. Quite the opposite, in fact. It pooled four different groups of people to get a range of diverse lifestyles, including Greeks, Anglo-Celtic Australians, and Swedish people, and it found that people who had completely different eating habits—and completely different lives, we might reasonably assume—also had different amounts of wrinkles.

To me this is not a great surprise, and it illustrates a very simple issue in epidemiological research called "confounding variables": these are things that are related both to the outcome you're measuring (wrinkles) and to the exposure you are measuring (food), but that you haven't thought of yet. They can confuse an apparently causal relationship, and you have to think of ways to exclude or minimize confounding variables to get to the right answer, or at least be very wary that they are there. In the case of this study, there are almost too many confounding variables to describe.

I eat well—with lots of olive oil, as it happens—and I don't have many wrinkles. I also have a middle-class background, plenty of money, an indoor job, and, if we discount infantile threats of litigation and violence from people who cannot tolerate any discussion of their ideas, a life largely free from strife. People with completely different lives will always have different diets, and different wrinkles. They will have different employment histories, different amounts of stress, different amounts of sun exposure, different levels of affluence, different levels of social support, different patterns of cosmetics use, and much more. I can imagine plenty of reasons why you might find that people who eat olive oil have fewer wrinkles, and the olive oil's having a causative role, an actual physical effect on your skin when you eat it, is fairly low down on my list.

Now, to be fair to nutritionists, they are not alone in failing to understand the importance of confounding variables, in their eagerness for a clear story. Every time you read in a newspaper that "moderate alcohol intake" is associated with some improved health outcome—less heart disease, less obesity, anything—to gales of delight from the alcohol industry, and, of course, from your friends, who say, "Ooh, well, you see, it's better for me to drink a little . . ." as they drink a lot—you are almost certainly witnessing a journalist of limited intellect, overinterpreting a study with huge confounding variables.

This is because—let's be honest here—teetotalers are not like everyone else. They will almost certainly have a reason for not drinking, and it might be moral, or cultural, or perhaps even medical, but there's a serious risk that whatever is causing them to be teetotal might also have other effects on their health, confusing the relationship between their drinking habits and their health outcomes. Like what? Well, perhaps people from specific ethnic groups who are teetotal are also more likely to be obese, so they are less healthy. Perhaps people who deny themselves the indulgence of alcohol are more likely to indulge in chocolate and chips. Perhaps preexisting ill health will force you to give up alcohol, and that's skewing the figures, making teetotalers look unhealthier than moderate drinkers. Perhaps these teetotalers are recovering alcoholics: among the people I know, they're the ones who are most likely to be absolute teetotalers, and they're also more likely to be fat, from all those years of heavy alcohol abuse. Perhaps some of the people who say they are teetotal are just lying.

This is why we are cautious about interpreting observational data, and to me Dowden has extrapolated too far from the data in her eagerness to dispense—with great authority and certainty— *very* specific dietary wisdom in her newspaper column (but of course, you may disagree, and you now have the tools to do so meaningfully).

If we were modern about this, and wanted to offer constructive criticism, what might she have written instead? I think, both here and elsewhere, that despite what journalists and self-appointed "experts" might say, people are perfectly capable of understanding the evidence for a claim, and anyone who withholds, overstates, or obscures that evidence, while implying that she's doing the reader a favor, is probably up to no good. MMR is an excellent parallel example of where the bluster, the panic, the "concerned experts," and the conspiracy theories of the media were very compelling, but the science itself was rarely explained.

So, leading by example, if I were a media nutritionist, I might say, if pushed, after giving all the other sensible sun advice, "A survey found that people who eat more olive oil have fewer wrinkles," and I might feel obliged to add, "Although people with different diets may differ in lots of other ways." But then, I'd also be writing about food, so: "Never mind, here's a delicious recipe for salad dressing anyway." Nobody's going to employ me to write a nutritionist column.

FROM THE LAB BENCH TO THE GLOSSIES

Nutritionists love to quote basic laboratory science research because it makes them look as if they are actively engaged in a process of complicated, impenetrable, highly technical academic work. But you have to be very cautious about how you extrapolate from what happens to some cells in a dish on a laboratory bench to the complex system of a living human being, where things can work in completely the opposite way from what laboratory work would suggest. Anything can kill cells in a test tube. Fairy liquid will kill cells in a test tube, but you don't take it to cure cancer. This is just another example of how nutritionism, despite the alternative medicine rhetoric and phrases like "holistic," is actually a crude, unsophisticated, old-fashioned, and, above all, *reductionist* tradition.

Here is another example from Michael van Straten—who has fallen sadly into our quadrat, and I don't want to introduce too many new characters or confuse you—writing in the British *Daily Express* as its nutrition specialist: "Recent research," he says, has shown that turmeric is "highly protective against many forms of cancer, especially of the prostate." It's an interesting idea, worth pursuing, and there have been some speculative lab studies of cells, usually from rats, growing or not growing under microscopes, with turmeric extract tipped on them. There is some limited animal model data, but it is not fair to say that turmeric, or curry, in the real world, in real people, is "highly protective against many forms of cancer, especially of the prostate," least of all because it's not very well absorbed, since your liver tends to absorb and metabolize it before it can reach other organs.

Forty years ago a man called Austin Bradford Hill, the grandfather of modern medical research, who was key in discovering the link between smoking and lung cancer, wrote out a set of guidelines, a kind of tick list, for assessing causality and a relationship between an exposure and an outcome. These are the cornerstone of evidence-based medicine, and often worth having at the back of your mind: it needs to be a strong association, which is consistent, and specific to the thing you are studying, where the putative cause comes before the supposed effect in time; ideally there should be a biological gradient, such as a dose-response effect; it should be consistent or at least not completely at odds with what is already known (because extraordinary claims require extraordinary evidence); and it should be biologically plausible.

Michael van Straten here has got biological plausibility, and little else. Medics and academics are very wary of people's making claims on such tenuous grounds, because it's something you get a lot from people with something to sell—specifically, drug companies.

Drug companies are very keen to promote theoretical advantages ("It works more on the Z4 receptor, so it must have fewer side effects!"), animal experiment data, or "surrogate outcomes" ("It improves blood test results; it must be protective against heart attacks!") as evidence of the efficacy or superiority of their products. Many of the more detailed popular nutritionist books, should you ever be lucky enough to read them, play this classic drug company card very assertively. They will claim, for example, that a "placebo-controlled randomized control trial" has shown *benefits* from a particular vitamin, when what they mean is, it showed changes in a "surrogate outcome."

For example, the trial may merely have shown that there were measurably increased amounts of the vitamin in the bloodstream after you took a vitamin, compared with placebo, which is a pretty unspectacular finding in itself; yet this is presented to the unsuspecting lay reader as a positive trial. Or the trial may have shown that there were changes in some other blood marker, perhaps the level of an ill-understood immune system component, which, again, the media nutritionist will present as concrete evidence of a real-world benefit.

There are problems with using such surrogate outcomes. They are often only tenuously associated with the real disease, in a very abstract theoretical model, and often developed in the very idealized world of an experimental animal, genetically inbred, kept under conditions of tight physiological control. A surrogate outcome can, of course, be used to generate and examine hypotheses about a real disease in a real person, but it needs to be very carefully validated. Does it show a clear dose-response relationship? Is it a true predictor of disease or merely a "covariable," something that is related to the disease in a different way (e.g., caused *by* it rather than involved in *causing* it)? Is there a well-defined cutoff between normal and abnormal values?

All I am doing, I should be clear, is taking the feted media nutritionists at their own word. They present themselves as men and women of science, fill their columns, TV shows, and books with references to scientific research. I am subjecting their claims to the exact same level of very basic, uncomplicated rigor that I would deploy for any new theoretical work, any drug company claim and pill marketing rhetoric, and so on.

It's not unreasonable to use surrogate outcome data, as they do, but those who are in the know are always circumspect. We're *interested* in early theoretical work, but often the message is: "It might be a bit more complicated than that . . ." You'd only want to accord a surrogate outcome any significance if you'd read everything on it yourself, or if you could be absolutely certain that the person assuring you of its validity was extremely capable and was giving a sound appraisal of all the research in a given field, and so on.

Similar problems arise with animal data. Nobody could deny that this kind of data is valuable in the theoretical domain, for developing hypotheses, or suggesting safety risks, when cautiously appraised. But media nutritionists, in their eagerness to make lifestyle claims, are all too often blind to the problems of applying these isolated theoretical nuggets to humans, and anyone would think they were just trawling the Internet looking for random bits of science to sell their pills and expertise (imagine that). Both the tissue and the disease in an animal model, after all, may be very different from those in a living human system, and these problems are even greater with a lab dish model. Giving unusually high doses of chemicals to animals can distort the usual metabolic pathways, and give misleading results, and so on. Just because something can upregulate or downregulate something in a model doesn't mean it will have the effect you expect in a person, as we shall see with the stunning truth about antioxidants.

And what about turmeric, which we were talking about before I tried to show you the entire world of applying theoretical research in this tiny grain of spice? Well, yes, there is some evidence that curcumin, a chemical in turmeric, is highly biologically active, in all kinds of different ways, on all kinds of different systems (there are also theoretical grounds for believing that it may be carcino-genic, mind you). It's certainly a valid target for research.

But for the claim that we should eat more curry in order to get more of it, that "recent research" has shown it is "highly protective against many forms of cancer, especially of the prostate," you might want to step back and put the theoretical claims in the con-text of your body. Very little of the curcumin you eat is absorbed. You have to eat a few grams of it to reach significant detectable serum levels, but to get a few grams of *curcumin*, you'd have to eat one hundred grams of *turmeric*, and good luck with that. Between research and recipe, there's a lot more to think about than the nutritionists might tell you.

CHERRY-PICKING

The idea is to try and give all the information to help others to judge the value of your contribution; not just the information that leads to judgment in one particular direction or another.

—Richard P. Feynman

There have been an estimated fifteen million medical academic articles published so far, and five thousand journals are published every month. Many of these articles will contain contradictory claims; picking out what's relevant—and what's not—is a gargan-tuan task. Inevitably people will take shortcuts. We rely on review

Cherry-Picking is my past education

articles, or on meta-analyses, or textbooks, or hearsay, or chatty journalistic reviews of a subject.

That's if your interest is in getting to the truth of the matter. What if you've just got a point to prove? There are few opinions so absurd that you couldn't find at least one person with a Ph.D. somewhere in the world to endorse them for you; and similarly, there are few propositions in medicine so ridiculous that you couldn't conjure up some kind of published experimental evidence somewhere to support them, if you didn't mind its being a tenuous relationship and cherry-picked the literature, quoting only the studies that were in your favor.

One of the great studies of cherry-picking in the academic literature comes from an article about Linus Pauling, the great-grandfather of modern nutritionism, and his seminal work on vitamin C and the common cold. In 1993 Paul Knipschild, professor of epidemiology at the University of Maastricht, published a chapter on Pauling in the mighty textbook *Systematic Reviews*; he had gone to the extraordinary trouble of approaching the literature as it stood when Pauling was working and subjecting it to the same rigorous systematic review that you would find in a modern paper.

He found that while some trials did suggest that vitamin C had some benefits, Pauling had selectively quoted from the literature to prove his point. Where Pauling had referred to some trials that seriously challenged his theory, it was to dismiss them as methodologically flawed; but as a cold examination showed, so too were papers he quoted favorably in support of his own case.

In Pauling's defense, his was an era when people knew no better, and he was probably quite unaware of what he was doing, but today cherry-picking is one of the most common dubious practices in alternative therapies, particularly in nutritionism, where it seems to be accepted essentially as normal practice (it is this cherry-picking, in reality, that helps characterize what alternative therapists conceive of, rather grandly, as their alternative paradigm). It happens in mainstream medicine also, but with one crucial difference: there it is recognized as a major problem, and hard work has been done to derive a solution.

That solution is a process called systematic review. Instead of just mooching around online and picking out your favorite papers to back up your prejudices and help you sell a product, in a systematic review you have an explicit search strategy for seeking out data (openly described in your paper, even including the search terms you used on databases of research papers), you tabulate the characteristics of each study you find, you measure—ideally blind to the results—the methodological quality of each one (to see how much of a "fair test" it is), you compare alternatives, and then finally you give a critical, weighted summary.

This is what the Cochrane Collaboration does on all the health care topics that it can find. It even invites people to submit new clinical questions that need answers. This careful sifting of information has revealed huge gaps in knowledge, it has revealed that "best practices" were sometimes murderously flawed, and simply by sifting methodically through preexisting data, it has saved more lives than you could possibly imagine. In the nineteenth century, as the public health doctor Muir Gray has said, we made great advances through the provision of clean, clear water; in the twenty-first century we will make the same advances through clean, clear information. Systematic reviews are one of the great ideas of modern thought. They should be celebrated.

PROBLEMATIZING ANTIOXIDANTS

We have seen the kinds of errors made by those in the nutritionism movement as they strive to justify their more obscure and technical claims. What's more fun is to take our new understanding and apply it to one of the key claims of the nutritionism movement, indeed to a fairly widespread belief in general: the claim that you should eat more antioxidants.

As you now know, there are lots of ways of deciding whether the totality of research evidence for a given claim stacks up, and it's rare that one single piece of information clinches it. In the case of a claim about food, for example, there are all kinds of different things we might look for: whether it is theoretically plausible, whether it is backed up by what we know from observing diets and health, whether it is supported by "intervention trials," in which we give one group one diet and another group a different one, and whether those trials measured real-world outcomes, like "death," or a surrogate outcome, like a blood test, which is only hypothetically related to a disease.

My aim here is by no means to suggest that antioxidants are *entirely* irrelevant to health. If I had a T-shirt slogan for this whole book, it would be: "I think you'll find it's a bit more complicated than that." I intend, as they say, to "problematize" the prevailing nutritionist view on antioxidants, which currently lags only about twenty years behind the research evidence.

From an entirely theoretical perspective, the idea that antioxidants are beneficial for health is an attractive one. When I was a medical student—not so long ago—the most popular biochemistry textbook was called *Stryer*. This enormous book is filled with complex interlocking flowcharts of how chemicals, which is what you are made of, move through the body. It shows how different enzymes break down food into its constituent molecular elements, how these are absorbed, how they are reassembled into new, larger molecules that your body needs to build muscles, retina, nerves, bone, hair, membrane, mucus, and everything else that you're made of; how the various forms of fats are broken down and reassembled into new forms of fat; or how different forms of molecule—sugar, fat, even alcohol—are broken down gradually, step by step, to release energy, and how that energy is transported, and how the incidental products from that process are used, or bolted onto something else to be transported in the blood, and then ditched at the kidneys, or metabolized down into further constituents, or turned into something useful elsewhere, and so on. This is one of the great miracles of life, and it is endlessly, beautifully, intricately fascinating.

When you look at these enormous, overwhelming interlocking webs, it's hard not to be struck by the versatility of the human body and how it can perform acts of near alchemy from so many different starting points. It would be very easy to pick one of the elements of these vast interlocking systems and become fixated on the idea that it is uniquely important. Perhaps it appears a lot on the diagram; or perhaps rarely, and seems to serve a uniquely important function in one key place. It would be easy to assume that if there were more of it around, then that function would be performed with greater efficiency.

But as with all enormous interlocking systems—societies, for example, or businesses—an intervention in one place can have quite unexpected consequences; there are feedback mechanisms, compensatory mechanisms. Rates of change in one localized area can be limited by quite unexpected factors that are entirely remote from what you are altering, and excesses of one thing in one place can distort the usual pathways and flows, to give counterintuitive results.

The theory underlying the view that antioxidants are good for you is the free radical theory of aging. Free radicals are highly chemically reactive, as are many things in the body. Often this reactivity is put to very good use. For example, if you have an infection, and there are some harmful bacteria in your body, then a phagocytic cell from your immune system might come along, identify the bacteria as unwelcome, build a strong wall around as many of them as it can find, and blast them with destructive free radicals. Free radicals are basically like bleach, and this process is a lot like pouring bleach down the toilet. Once again, the human body is cleverer than anybody you know.

But free radicals in the wrong places can damage the desirable components of cells. They can damage the lining of your arteries, and they can damage DNA, and damaged DNA leads to aging or cancer, and so on. For this reason, it has been suggested that free radicals are responsible for aging and various diseases. This is a theory, and it may or may not be correct.

Antioxidants are compounds which can—and do—"mop up" these free radicals, by reacting with them. If you look at the vast, interlocking flowchart diagrams of how all the molecules in your body are metabolized from one form to the next, you can see that this is happening all over the shop.

The theory that antioxidants are protective is separate from, but builds upon, the free radical theory of disease. If free radicals are dangerous, the argument goes, and antioxidants on the big diagrams are involved in neutralizing them, then eating more antioxidants should be good for you and reverse or slow aging and prevent disease.

There are a number of problems with this as a theory. First, who says free radicals are always bad? If you're going to reason just from theory, and from the diagrams, then you can hook all kinds of things together and make it seem as if you're talking sense. As I said, free radicals are vital for your body to kill off bacteria in phagocytic immune cells, so should you set yourself up in business and market an antioxidant-*free* diet for people with bacterial infections?

Second, just because antioxidants are involved in doing something good, why should eating more of them necessarily make that process more efficient? I know it makes sense superficially; but so do a lot of things, and that's what's really interesting about science (and this story in particular): sometimes the results aren't quite what you might expect. Perhaps an excess of antioxidants is simply excreted or turned into something else. Perhaps it just sits there doing nothing, because it's not needed. After all, half a tank of gasoline will get you across town just as easily as a full tank. Or perhaps, if you have an unusually enormous amount of antioxidant lying around in your body doing nothing, it doesn't just do nothing. Perhaps it does something actively harmful. That would be a turnup for the books, wouldn't it?

There were a couple of other reasons why the antioxidant theory seemed like a good idea twenty years ago. First, when you take a static picture of society, people who eat lots of fresh fruits and vegetables tend to live longer and have less cancer and heart disease, and there are lots of antioxidants in fruit and vegetables (although there are lots of other things in them too, and, you might rightly assume, lots of other healthy things about the lives of people who eat lots of healthy fresh fruit and vegetables, like their posh jobs, moderate alcohol intake, etc.).

Similarly, when you take a snapshot picture of the people who take antioxidant supplement pills, you will often find that they are healthier or live longer: but again (although nutritionists are keen to ignore this fact), these are simply surveys of people who have already chosen to take vitamin pills. These are people who are more likely to care about their health and are different from the everyday population—and perhaps from you—in lots of other ways, far beyond their vitamin pill consumption: they may take more exercise, have more social supports, smoke less, drink less, and so on.

But the early evidence in favor of antioxidants was genuinely promising and went beyond mere observational data on nutrition and health; there were also some very seductive blood results. In 1981 Richard Peto, one of the most famous epidemiologists in the world, who shares the credit for discovering that smoking causes 95 percent of lung cancer, published a major paper in *Nature*. He reviewed a number of studies that apparently showed a positive relationship between having a lot of beta-carotene on board (this is an antioxidant available in the diet) and a reduced risk of cancer.

This evidence included case-control studies, in which people *with* various cancers were compared with people *without* cancer (but matched for age, social class, gender, and so on), and it was found that the cancer-free subjects had higher plasma carotene. There were also prospective cohort studies, in which people were classified by their plasma carotene level at the beginning of the study, before any of them had cancer, and then followed up for many years. These studies showed twice as much lung cancer in the group with the lowest plasma carotene, compared with those with the highest level. It looked as if having more of these anti-oxidants might be a very good thing.

Similar studies showed that higher plasma levels of antioxidant vitamin E were related to lower levels of heart disease. It was suggested that vitamin E status explained much of the variations in levels of ischemic heart disease between different countries in Europe, which could not be explained by differences in plasma cholesterol or blood pressure.

But the editor of *Nature* was cautious. A footnote was put onto the Peto paper that read as follows: "Unwary readers (if such there are) should not take the accompanying article as a sign that the consumption of large quantities of carrots (or other dietary sources of beta-carotene) is necessarily protective against cancer." It was a very prescient footnote indeed.

THE ANTIOXIDANT DREAM UNRAVELS

Whatever the shrill alternative therapists may say, doctors and academics have an interest in chasing hints that could bear fruit, and compelling hypotheses like these, which could save millions of lives, are not taken lightly. These studies were acted upon, with many huge trials of vitamins set up and run around the world. There's also an important cultural context for this rush of activity that cannot be ignored: it was the tail end of the golden age of medicine. Before 1935 there weren't too many effective treatments around: we had insulin, liver for iron-deficiency anemia, and morphine— a drug with superficial charm at least—but in many respects, doctors were fairly useless. Then suddenly, between about 1935 and 1975, science poured out a constant stream of miracles.

Almost everything we associate with modern medicine happened in that time: treatments like antibiotics, dialysis, transplants, intensive care, heart surgery, almost every drug you've ever heard of, and more. As well as the miracle treatments, we really were finding those simple, direct, hidden killers that the media still pine for so desperately in their headlines. Smoking, to everybody's genuine surprise—one single risk factor—turned out to cause almost all lung cancer. And asbestos, through some genuinely brave and subversive investigative work, was shown to cause mesothelioma.

The epidemiologists of the 1980s were on a roll, and they believed that they were going to find lifestyle causes for all the major diseases of humankind. A discipline that had got cracking when John Snow took the handle off the Broad Street pump in 1854, terminating that pocket of the Soho cholera epidemic by cutting off the supply of contaminated water (it was a bit more complicated than that, but we don't have the time here), was going to come into its own. They were going to identify more and more of these one-to-one correlations between exposures and disease, and in their fervent imaginations, with simple interventions and cautionary advice they were going to save whole nations of people. This dream was very much not realized, as it turned out to be a bit more complicated than that.

Two large trials of antioxidants were set up after Peto's paper (which rather gives the lie to nutritionists' claims that vitamins are never studied because they cannot be patented: in fact, there have been a great many such trials, although the food supplement industry, estimated by one report to be worth over fifty billion dollars globally, rarely deigns to fund them). One was in Finland, where

thirty thousand participants at high risk of lung cancer were recruited, and randomized to receive beta-carotene, vitamin E, or
both or neither. Not only were there more lung cancers among the
people receiving the supposedly protective beta-carotene supplements, compared with placebo, but this vitamin group also had
more deaths overall, from both lung cancer and heart disease.

The results of the other trial were almost worse. It was called
the Carotene and Retinol Efficacy Trial, or CARET, in honor of the
high carotene content of carrots. It's interesting to note, while we're
here, that carrots were the source of one of the great disinformation
coups of World War II, when the Germans couldn't understand how
our pilots could see their planes coming from huge distances, even
in the dark. To stop them from trying to work out if we'd invented
anything clever like radar (as we had), the British instead started an
elaborate and entirely made-up nutritionist rumor. Carotenes in
carrots, they explained, are transported to the eye and converted to
retinal, which is the molecule that detects light in the eye (this is
basically true and is a plausible mechanism, like those we've already
dealt with), so, went the story, doubtless with much chortling behind their excellent RAF mustaches, we have been feeding our
chaps huge plates of carrots, to jolly good effect.

Anyway. Two groups of people at high risk of lung cancer were
studied: smokers, and people who had been exposed to asbestos at
work. Half were given beta-carotene and vitamin A, while the
other half got placebo. Eighteen thousand participants were due
to be recruited throughout its course, and the intention was that
they would be followed up for an average of six years; but in fact,
the trial was terminated early, because it was considered unethical
to continue it. Why? The people having the antioxidant tablets
were 46 percent more likely to die from lung cancer, and 17 percent more likely to die of any cause,* than the people taking pla-

*I have deliberately expressed this risk in terms of the "relative risk increase," as part of a
dubious in joke with myself. You will learn about this starting on page 186.

cebo pills. This is not news, hot off the presses; it happened well over a decade ago.

Since then the placebo-controlled trial data on antioxidant vitamin supplements has continued to give negative results. The most up-to-date Cochrane reviews of the literature pool together all the trials on the subject, after sourcing the widest possible range of data using the systematic search strategies described above (rather than cherry-picking studies to an agenda); they assess the quality of the studies and then put them all into one giant spreadsheet to give the most accurate possible estimate of the risks of benefits, and they show that antioxidant supplements are either ineffective or perhaps even actively harmful.

The Cochrane review on preventing lung cancer pooled data from four trials, describing the experiences of more than one hundred thousand participants and found no benefit from antioxidants, and indeed an increase in risk of lung cancer in participants taking beta-carotene and retinol together. The most up-to-date systematic review and meta-analysis on the use of antioxidants to reduce heart attacks and stroke looked at vitamin E, and separately beta-carotene, in fifteen trials, and found no benefit for either. For beta-carotene, there was a small but significant increase in death.

Most recently, a Cochrane review looked at the number of deaths, from any cause, in all the placebo-controlled randomized trials on antioxidants that have ever been performed (many of which looked at quite high doses, but perfectly in line with what you can buy in health food stores), describing the experiences of 230,000 people in total. This showed that overall, antioxidant vitamin pills do not reduce deaths, and in fact, they may increase your chance of dying.

Where does all this leave us? There was an observed correlation between low blood levels of these antioxidant nutrients and a higher incidence of cancer and heart disease, and a plausible mechanism for how they could have been preventive, but when you gave them as supplements, it turned out that people were no better off or were possibly *more* likely to die. That is, in some respects, a shame, as nice quick fixes are always useful, but there you go. It means that something funny is going on, and it will be interesting to get to the bottom of it and find out what.

More interesting is how uncommon it is for people even to be aware of these findings about antioxidants. There are various reasons why this has happened. First, it's an unexpected finding, although in that regard antioxidants are hardly an isolated case. Things that work in theory often do not work in practice, and in such cases we need to revise our theories, even if it is painful. Hormone replacement therapy seemed like a good idea for many decades, until the follow-up studies revealed the problems with it, so we changed our views. And calcium supplements once looked like a good idea for osteoporosis, but now it turns out that they probably increase the risk of heart attacks in older women, so we change our view.

It's a chilling thought that when we think we are doing good, we may actually be doing harm, but it is one we must always be alive to, even in the most innocuous situations. The pediatrician Dr. Benjamin Spock wrote a record-breaking bestseller titled *Baby and Child Care*, first published in 1946, that was hugely influential and largely sensible. In it, he confidently recommended that babies should sleep on their tummies. Dr. Spock had little to go on; but we now know that this advice is wrong, and the apparently trivial suggestion contained in his book, which was so widely read and followed, has led to thousands, and perhaps even tens of thousands, of avoidable crib deaths. The more people are listening to you, the greater the effects of a small error can be. I find this simple anecdote deeply disturbing.

But of course, there is a more mundane reason why people may not be aware of these findings on antioxidants, or at least may not take them seriously, and that is the phenomenal lobbying power of a large, sometimes rather dirty industry, which sells a lifestyle product that many people feel passionately about. The food supplement industry has engineered itself a beneficent public image, but this is not borne out by the facts. First, there is essentially no difference between the vitamin industry and the pharmaceutical and biotech industries (that is one message of this book, after all: the tricks of the trade are the same the world over). Key players include companies like Roche and Sanofi-Aventis; BioCare, the U.K. vitamin pill company, is part owned by Elder Pharmaceuticals, and so on. The vitamin industry is also—amusingly—legendary in the world of economics as the setting of the most outrageous price-fixing cartel ever documented. During the 1990s the main offenders were forced to pay *the largest criminal fines ever levied in legal history*—$1.5 billion in total—after entering guilty pleas with the U.S. Department of Justice and regulators in Canada, Australia, and the European Union. That's quite some cozy cottage industry.

Whenever a piece of evidence is published suggesting that the fifty-billion-dollar food supplement industry's products are ineffective, or even harmful, an enormous marketing machine lumbers into life, producing spurious and groundless methodological criticisms of the published data in order to muddy the waters—not enough to be noteworthy in a meaningful academic discussion, but that is not their purpose. This is a well-worn risk management tactic from many industries, including those producing tobacco, asbestos, lead, vinyl chloride, chromium, and more. It is called manufacturing doubt, and in 1969 one tobacco executive was stupid enough to commit it to paper in a memo. "Doubt is our product," he wrote, "since it is the best means of competing with the 'body of fact' that exists in the minds of the general public. It is also the means of establishing a controversy."

Nobody in the media dares challenge these tactics, where lobbyists raise sciencey-sounding defenses of their products, because they feel intimidated and lack the skills to do so. Even if they did, there would simply be a confusing and technical discussion on the radio, which everyone would switch off, and at most the consumer would hear only "controversy": job done. I don't think that food supplement pills are as dangerous as tobacco—few things are—but it's hard to think of any other kind of pill for which research could be published showing a possible increase in death, and industry representatives would be wheeled out and given as easy a ride as the vitamin companies' employees are given when papers are published on their risks. But then, of course, many of them have their own slots in the media to sell their wares and their worldview.

The antioxidant story is an excellent example of how wary we should be of blindly following hunches based on laboratory-level and theoretical data, and naively assuming, in a reductionist manner, that this must automatically map onto dietary and supplement advice, as the media nutritionists would have us do. It is an object lesson in what an unreliable source of research information these characters can be, and we would all do well to remember this story the next time someone tries to persuade us with blood test data, or talk about molecules, or theories based on vast, interlocking metabolism diagrams that we should buy his book, her wacky diet, or his bottle of pills.

More than anything it illustrates how this atomized, overcomplicated view of diet can be used to mislead and oversell. I don't think it's melodramatic to speak of people disempowered and paralyzed by confusion, with all the unnecessarily complex and conflicting messages about food. If you're really worried, you can buy Fruitella Plus, which are nasty, chewy sweets but with added vitamins A, C, E, and calcium. In the last five years, even chocolate companies have jumped on the bandwagon, in the ultimate expression of how nutritionism has perverted and distorted our common sense about food.

Mars unveiled CocoaVia, a chocolate with extra antioxidants that they claimed was good for hearts and arteries. Hershey's soon followed with "Natural Flavanol Antioxidant Milk Chocolate," as part of its healthy "Goodness Line."

If I were writing a lifestyle book, it would have the same advice on every page, and you'd know it all already. Eat lots of fruit and vegetables, and live your whole life in every way as well as you can: exercise regularly as part of your daily routine, avoid obesity, don't drink too much, don't smoke, and don't get distracted from the real, basic, simple causes of ill health. But as we shall see, even these things are hard to do on your own and in reality require wholesale social and political changes.

7

NUTRITIONISTS

So who are these people? The most important thing to recognize is that there is nothing new here. Although the contemporary nutritionism movement likes to present itself as a thoroughly modern and evidence-based enterprise, the food guru industry, with its outlandish promises, moralizing, and sexual obsessions, goes back at least two centuries.

Like our modern food gurus, the historical figures of nutritionism were mostly enthusiastic laypeople, and they all claimed to understand nutritional science, evidence, and medicine better than the scientists and doctors of their era. The advice and the products may have shifted with prevailing religious and moral notions, but they have always played to the market, be it puritan or liberal, New Age or Christian.

Graham crackers are a digestive biscuit invented in the nineteenth century by Sylvester Graham, the first great advocate of vegetarianism and nutritionism as we would know it, and proprietor of the world's first health food shop. Like his descendants today, Graham mixed up sensible notions, such as cutting down on cigarettes and alcohol, with some other, rather more esoteric, ideas that he concocted for himself. He warned that ketchup and mustard, for example, can cause "insanity."

I've got no great beef with the organic food movement (even if its claims are a little unrealistic), but it's still interesting to note that Graham's health food store—in 1837—heavily promoted its food as being grown according to "physiological principles" on "virgin unvitiated soil." By the retro-fetishism of the time, this was soil that had not been "subjected" to "overstimulation" . . . by manure.

Soon these food marketing techniques were picked up by more overtly puritanical religious zealots like John Harvey Kellogg, one of the men behind the cornflake. Kellogg was a natural healer and health food advocate, promoting his granola bars as the route to abstinence, temperance, and solid morals. He ran a sanatorium for private clients, using "holistic" techniques, including that modern favorite colonic irrigation.

Kellogg was also a keen antimasturbation campaigner. He advocated exposing the tissue on the end of the penis, so that it smarted with friction during acts of self-pollution (and you do have to wonder about the motives of anyone who thinks the problem through in that much detail). Here is a particularly enjoyable passage from his *Treatment for Self-Abuse and Its Effects* (1888), in which Kellogg outlines his views on circumcision: "The operation should be performed by a surgeon without administering an anesthetic, as the brief pain attending the operation will have a salutary effect upon the mind, especially if it be connected with the idea of punishment. In females, the author has found the application of pure carbolic acid to the clitoris an excellent means of allaying the abnormal excitement."

By the early twentieth century a man named Bernard Macfadden had updated the nutritionism model for contemporary moral values and so became the most commercially successful health guru of his time. He changed his Christian name from Bernard to Bernarr, because it sounded more like the roar of a lion (this is completely true), and ran a successful magazine titled *Physical Culture*, featuring beautiful bodies doing healthy things. The pseudoscience and the posturing were the same, but he used liberal sexuality to his advantage, selling his granola bars as a food that would promote a muscular, thrusting, lustful lifestyle in that decadent rush that flooded the populations of the West between the wars.*

More recently there was Dudley J. LeBlanc, a Louisiana senator and the man behind Hadacol ("I hadda call it something"). It cured everything, cost $100 a year for the recommended dose, and to Dudley's open amazement, it sold in the millions. "They came in to buy Hadacol," said one pharmacist, "when they didn't have money to buy food. They had holes in their shoes and they paid $3.50 for a bottle of Hadacol."

LeBlanc made no medicinal claims, but pushed customer testimonials to an eager media. He appointed a medical director who had been convicted in California of practicing medicine with no license and no medical degree. A diabetic patient almost died when she gave up insulin to treat herself with Hadacol, but nobody cared. "It's a craze. It's a culture. It's a political movement," said *Newsweek*.

It's easy to underestimate the phenomenal and enduring commercial appeal of these kinds of products and claims throughout

---

*Interestingly, Macfadden's food product range was complemented by a more unusual invention of his own. The Peniscope was a popular suction device designed to enlarge the male organ that is still used by many today, in a modestly updated form. Since this may be your only opportunity to learn about the data on penis enlargement, it's worth mentioning that there is in fact some evidence that stretching devices can increase penis size. Gillian McKeith's Wild Pink and Horny Goat Weed sex supplement pills, however, sold for "maintaining erections, orgasmic pleasure, ejaculation . . . lubrication, satisfaction, and arousal," could claim no such evidence for efficacy (and in 2007, after much complaining, these seedy and rather old-fashioned products were declared illegal by the Medicines and Healthcare Products Regulatory Agency, or MHRA). I mention this only because, rather charmingly, it means that Macfadden's Peniscope may have a better evidence base for its claims than either his own food products or McKeith's Horny Goat penis pills.

history. By 1950 Hadacol's sales were over twenty million dollars with an advertising spend of one million dollars a month, in 700 daily papers and on 528 radio stations. LeBlanc took a traveling medicine show of 130 vehicles on a tour of thirty-eight hundred miles through the South. Entry was paid in Hadacol bottle tops, and the shows starred Groucho and Chico Marx, Mickey Rooney, Judy Garland, and educational exhibitions of scantily clad women illustrating "the history of the bathing suit." Dixieland bands played songs like "Hadacol Boogie" and "Who Put the Pep in Grandma?"

The senator used Hadacol's success to drive his political career, and his competitors, the Longs—descended from the Democrat reformer Huey Long—panicked, launching their own patent medicine called Vita-Long. By 1951 LeBlanc was spending more in advertising than he was making in sales, and in February of that year, shortly after he sold the company—and shortly before it folded—he appeared on the TV show *You Bet Your Life* with his old friend Groucho Marx. "Hadacol," said Groucho, "what's that good for?" "Well," said LeBlanc, "it was good for about five and a half million dollars for me last year." The point I am making is that there is nothing new under the sun. There have always been health gurus selling magic potions, and there always will be.

Let's look at just one: Dr. Gillian McKeith—a prime-time TV celebrity in the U.K., now a rising star on BBC America, and a bestselling author with an empire of products. To some she is a guru. To me she is, as we shall see, a menace to the public understanding of science. She has a mainstream television nutrition show, yet she seems to misunderstand not nuances but the most basic aspects of biology, things that a schoolchild could put her straight on.

I first noticed Dr. Gillian McKeith when a reader sent in a clipping about her first series on Channel 4. McKeith was styled, very strikingly, as a white-coated academic and scientific authority on nutrition, a "clinical nutritionist," posing in laboratories, surrounded by test tubes, and talking about diagnoses and molecules.

She was also quoted here saying something a fourteen-year-old doing GCSE biology could easily have identified as pure nonsense: recommending spinach and the darker leaves on plants, because they contain more chlorophyll. According to McKeith, these are "high in oxygen" and will "really oxygenate your blood." This same claim is repeated all over her books.

Forgive me for patronizing, but before we go on, you may need a little refresher on the miracle of photosynthesis. Chlorophyll is a small green molecule that is found in chloroplasts, the miniature factories in plant cells that take the energy from sunlight and use it to convert carbon dioxide and water into sugar and oxygen. Using this process, called photosynthesis, plants store the energy from sunlight in the form of sugar (high in calories, as you know), and they can then use this sugar energy to make everything else they need, like protein, and fiber, and flowers, and corn on the cob, and bark, and leaves, and amazing traps that eat flies, and cures for cancer, and tomatoes, and wispy dandelions, and conkers, and chilies, and all the other amazing things that the plant world has going on.

Meanwhile, you breathe in the oxygen that the plants give off during this process—essentially as a by-product of their sugar manufacturing—and you also eat the plants, or you eat animals that eat the plants, or you build houses out of wood, or you make painkiller from willow bark, or any of the other amazing things that happen with plants. You also breathe out carbon dioxide, and the plants can combine that with water to make more sugar again, using the energy from sunlight, and so the cycle continues.

Like most things in the story the natural sciences can tell about the world, it's all so beautiful, so gracefully simple, yet so rewardingly complex, so neatly connected—not to mention true—that I can't even begin to imagine why anyone would ever want to believe some New Age "alternative" nonsense instead. I would go so far as to say that even if we all are under the control of a benevo-

lent God, and the whole of reality turns out to come down to some
flaky spiritual "energy" that only alternative therapists can truly
harness, that's still neither so interesting nor so graceful as the most
basic stuff I was taught at school about how plants work.

Is chlorophyll "high in oxygen"? No. It helps make oxygen. In
sunlight. And it's pretty dark in your bowels; in fact, if there's any
light in there at all, something's gone badly wrong. So any chlorophyll you eat will not create oxygen, and even if it did, even if
Dr. Gillian McKeith, Ph.D., stuck a searchlight right up your bum
to prove her point, and your salad began photosynthesizing, even
if she insufflated your guts with carbon dioxide through a tube, to
give the chloroplasts something to work with, and by some miracle you really did start to produce oxygen in there, you still wouldn't
absorb a significant amount of it through your bowel, because your
bowel is adapted to absorb food, while your lungs are optimized to
absorb oxygen. You do not have gills in your bowels. Neither, since
we've mentioned them, do fish. And while we're talking about it,
you probably don't want oxygen inside your abdomen anyway. In
keyhole surgery, surgeons have to inflate your abdomen to help
them see what they're doing, but they don't use oxygen, because
there's methane fart gas in there too, and we don't want anyone
catching fire on the inside. [There is no oxygen in your bowel.]

So who is this person, and how did she come to be teaching us
about diet? What possible kind of science degree can she have, to
be making such basic mistakes that a schoolkid would spot? Was
this an isolated error? A one-off slip of the tongue? I think not.

Actually, I know not, because as soon as I saw that ridiculous
quote, I ordered some more McKeith books. Not only does she
make the same mistake in numerous other places, but it seems to
me that her understanding of even the most basic elements of science is deeply, strangely distorted. In *You Are What You Eat* (page
211) she says: "Each sprouting seed is packed with the nutritional
energy needed to create a full grown healthy plant."

This is hard to follow. Does a fully grown, healthy oak tree, a hundred feet tall, contain the same amount of energy as a tiny acorn? No. Does a fully grown, healthy sugarcane plant contain the same amount of nutritional energy—measure it in "calories" if you like—as a sugarcane seed? No. Stop me if I'm boring you, in fact, stop me if I've misunderstood something in what she's said, but to me this seems like almost the same mistake as the photosynthesis thing, because that extra energy to grow a fully grown plant comes, again, from photosynthesis, in which plants use light to turn carbon dioxide and water into sugar and then into everything else that plants are made of.

This is not an incidental issue, an obscure backwater of McKeith's work, nor is it a question of which "school of thought" you speak for: the "nutritional energy" of a piece of food is one of the most important things you could possibly think of for a nutritionist to know about. I can tell you for a fact that the amount of nutritional energy you will get from eating one sugarcane seed is a hell of a lot less than you'd get from eating all the sugarcane from the plant that grew from it. These aren't passing errors or slips of the tongue (I have a policy, as it were, of not quibbling on spontaneous utterances, because we all deserve the chance to fluff); these are clear statements from published tomes.

If you watch McKeith's TV show with the eye of a doctor, it rapidly becomes clear that even here, frighteningly, she doesn't seem to know what she's talking about. She examines patients' abdomens on an examination couch as if she were a doctor and confidently announces that she can feel which organs are inflamed. But clinical examination is a fine art at the best of times, and what she is claiming is like identifying which fluffy toy someone has hidden under a mattress (you're welcome to try this at home).

She claims to be able to identify lymphedema, swollen ankles from fluid retention, and she almost does it right; at least, she kind of puts her fingers in roughly the right place, but only for about half a second, before triumphantly announcing her findings. If you'd like to borrow my second edition copy of Epstein and de Bono's *Clinical Examination* (I don't think there were many people in my year at medical school who didn't buy a copy), you'll discover that to examine for lymphedema, you press firmly for around thirty seconds, to gently compress the exuded fluid out of the tissues, then take your fingers away and look to see if they have left a dent behind.

In case you think I'm being selective, and quoting only McKeith's most ridiculous moments, there's more: the tongue is "a window to the organs—the right side shows what the gallbladder is up to, and the left side the liver." Raised capillaries on your face are a sign of "digestive enzyme insufficiency—your body is screaming for food enzymes." Thankfully, Gillian can sell you some food enzymes from her website. "Skid mark stools" (she is obsessed with feces and colonic irrigation) are "a sign of dampness inside the body. If your stools are foul-smelling, you are "sorely in need of digestive enzymes." Again. Her treatment for pimples on the forehead—not pimples anywhere else, mind you, only on the forehead—is a regular enema. Cloudy urine is "a sign that your body is damp and acidic, due to eating the wrong foods." The spleen is "your energy battery."

So we have seen scientific facts—on which Dr. McKeith seems to be mistaken. What of scientific process? She has claimed, repeatedly and to anyone who will listen, that she is engaged in clinical scientific research. Let's step back a moment, because from everything I've said, you might reasonably assume that McKeith has been clearly branded as some kind of alternative therapy maverick. In fact, nothing could be further from the truth. This doctor has been presented, consistently, up front on television, on her website, by her management company and in her books, as a scientific authority on nutrition.

Many watching her TV show quite naturally assumed she was a medical doctor. And why not? There she was, examining patients, performing and interpreting blood tests, wearing a white coat, surrounded by test tubes, "Dr. McKeith," "the diet doctor," giving diagnoses, talking authoritatively about treatment, using complex scientific terminology with all the authority she could muster, and sticking irrigation equipment nice and invasively right up into people's rectums.

Now, to be fair, I should mention something about the doctorate, but I should also be clear: I don't think this is the most important part of the story. It's the funniest and most memorable part of the story, but the real action is whether McKeith is capable of truly behaving like the nutritional science academic she claims to be.

And the scholarliness of her work is a thing to behold. She produces lengthy documents that have an air of "referenciness," with nice little superscript numbers, that talk about trials, and studies, and research, and papers . . . but when you follow the numbers, and check the references, it's shocking how often they aren't what she claimed them to be in the main body of the text, or they refer to funny little magazines and books, such as *Delicious, Creative Living, Healthy Eating*, and my favorite, *Spiritual Nutrition and the Rainbow Diet*, rather than proper academic journals.

She even does this in the book *Miracle Superfood*, which, we are told, is the published form of her Ph.D. "In laboratory experiments with anemic animals, red-blood cell counts have returned to normal within four or five days when chlorophyll was given," she says. Her reference for this experimental data is a magazine titled *Health Store News*. "In the heart," she explains, "chlorophyll aids in the transmission of nerve impulses that control contraction," a statement that is referenced to the second issue of a magazine titled *Earthletter*. Fair enough, if that's what you want to read—I'm bending over to be reasonable here—but it's clearly not a suitable source to reference that claim. This is her Ph.D., remember.

To me this is cargo cult science, as Professor Richard Feynman described it more than thirty years ago, in reference to the similarities between pseudoscientists and the religious activities on a few small Melanesian islands in the 1950s:

During the war they saw aeroplanes with lots of good materials, and they want the same thing to happen now. So they've arranged to make things like runways, to put fires along the sides of the runways, to make a wooden hut for a man to sit in, with two wooden pieces on his head as headphones and bars of bamboo sticking out like antennas—he's the controller—and they wait for the aeroplanes to land. They're doing everything right. The form is perfect. It looks exactly the way it looked before. But it doesn't work. No aeroplanes land.

Like the rituals of the cargo cult, the form of McKeith's pseudo-academic work is superficially correct: the superscript numbers are there, the technical words are scattered about, she talks about research and trials and findings; but the substance is lacking. I actually don't find this very funny. It makes me quite depressed to think about her, sitting up, perhaps alone, studiously and earnestly typing this stuff out.

McKeith's Ph.D. is from Clayton College of Natural Health, a nonaccredited correspondence course college, which, unusual for an academic institution, also sells its own range of vitamin pills through its website. Her master's degree is from the same august institution. At current Clayton prices, it's $6,400 in fees for the Ph.D., and less for the master's, but if you pay for both at once you get a $300 discount (and if you really want to push the boat out, Clayton has a package deal: two doctorates and a master's for $12,100 all in).

On her CV, posted on her management website, McKeith claimed to have a Ph.D. from the rather good American College of Nutrition. When this was pointed out, her representative explained that this was merely a mistake, made by a Spanish work experience kid who posted the wrong CV. The attentive reader may have noticed that the very same claim about the American College of Nutrition was also in one of her books from several years previously.

In 2007 a regular from my website—I could barely contain my pride—took McKeith to the Advertising Standards Authority, complaining about her using the title "doctor" on the basis of a qualification gained by correspondence course from a nonaccredited American college, and won. The ASA came to the view that McKeith's advertising breached two clauses of the Committee of Advertising Practice code: "substantiation" and "truthfulness."

Dr. McKeith sidestepped the publication of a damning ASA draft adjudication at the last minute by accepting—"voluntarily"—not to call herself doctor in her advertising anymore. In the news coverage that followed, McKeith suggested that the adjudication was concerned only with whether she had presented herself as a medical doctor. Again, this is not true. A copy of that draft adjudication has fallen into my lap—imagine that—and it specifically says that people seeing the ads would reasonably expect her to have either a medical degree or a Ph.D. from an accredited university.

She even managed to get one of her corrections into a profile on her in my own newspaper, *The Guardian*:

> Doubt has also been cast on the value of McKeith's certified membership of the American Association of Nutritional Consultants, especially since *Guardian* journalist Ben Goldacre managed to buy the same membership online for his dead cat for $60. McKeith's spokeswoman says of this membership: "Gillian has 'professional membership,' which is membership designed for practicing nutritional and dietary professionals, and is distinct from 'associate membership,' which is open to all individuals. To gain professional membership Gillian provided proof of her degree and three professional references.'"

Well. My dead cat Hettie is also a "certified professional member" of the American Association of Nutritional Consultants. I have the certificate hanging in my bathroom. Perhaps it didn't even occur to the journalist that McKeith could be wrong. More likely, in the tradition of nervous journalists, I suspect that she was hurried, on deadline, and felt she had to get McKeith's "right of reply" in, even if it cast doubts on—I'll admit my beef here—my own hard-won investigative revelations about my dead cat. I mean, I don't sign my dead cat up to bogus professional organizations for the good of my health, you know. It may sound disproportionate to suggest that I will continue to point out these obfuscations for as long as they are made, but I will, because to me there is a strange fascination in tracking their true extent.

If you contact the Australasian College of Health Sciences (Portland, Oregon), where McKeith has a "pending diploma in herbal medicine," it says it can't tell you anything about its students. If you contact Clayton College of Natural Health to ask where you can read her Ph.D. it says you can't. What kinds of organizations are these? If I said I had a Ph.D. from Cambridge, U.S. or U.K. (I have neither, and I claim no great authority), it would take you only a day to find it in their library.

For me the most concerning aspect of the way she responds to questioning of her scientific ideas is exemplified by a story from 2000, when Dr. McKeith approached a retired professor of nutritional medicine from the University of London. Shortly after the publication of her book *Living Food for Health*, John Garrow wrote an article about some of the bizarre scientific claims Dr. McKeith was making, and his piece was published in a fairly obscure medical newsletter. He was struck by the strength with which she presented her credentials as a scientist ("I continue every day to research, test and write furiously so that you may benefit . . ." etc). He has since said that he assumed—like many others—that she was a proper doctor. Sorry: a medical doctor. Sorry: a qualified, conventional medical doctor who has attended an accredited medical school.

In this book McKeith promised to explain how you can "boost your energy, heal your organs and cells, detoxify your body, strengthen your kidneys, improve your digestion, strengthen your immune system, reduce cholesterol and high blood pressure, break down fat, cellulose and starch, activate the enzyme energies of your body, strengthen your spleen and liver function, increase mental and physical endurance, regulate your blood sugar, and lessen hunger cravings and lose weight."

These are not modest goals, but her thesis was that they all were possible with a diet rich in enzymes from "live" raw food— fruits, vegetables, seeds, nuts, and especially live sprouts, which are "the food sources of digestive enzymes." She even offered "combination living food powder for clinical purposes," in case people didn't want to change their diets, and explained that she used this for "clinical trials" with patients at her clinic.

Garrow was skeptical of her claims. Apart from anything else, as emeritus professor of human nutrition at the University of London, he knew that human animals have their own digestive enzymes, and any plant enzyme you eat is likely to be digested like any other protein. As any professor of nutrition, and indeed many high school biology students, could tell you.

Garrow read McKeith's book closely, as have I. These "clinical trials" seemed to be a few anecdotes about how incredibly well her patients felt after seeing her. No controls, no placebo, no attempt to quantify or measure improvements. So Garrow made a modest proposal in a fairly obscure medical newsletter. I am quoting it in its entirety, partly because it is a rather elegantly written exposition of the scientific method by an eminent academic authority on the science of nutrition, but mainly because I want you to see how politely he stated his case:

> I also am a clinical nutritionist, and I believe that many of the statements in this book are wrong. My hypothesis is that any benefits which Dr. McKeith has observed in her

patients who take her living food powder have nothing to do with their enzyme content. If I am correct, then patients given powder which has been heated above 118°F for twenty minutes will do just as well as patients given the active powder. This amount of heat would destroy all enzymes, but make little change to other nutrients apart from vitamin C, so both groups of patients should receive a small supplement of vitamin C (say 60mg/day). However, if Dr. McKeith is correct, it should be easy to deduce from the boosting of energy, etc., which patients received the active powder and which the inactivated one.

Here, then, is a testable hypothesis by which nutritional science might be advanced. I hope that Dr. McKeith's instincts, as a fellow-scientist, will impel her to accept this challenge. As a further inducement I suggest we each post, say, £1,000, with an independent stakeholder. If we carry out the test, and I am proved wrong, she will of course collect my stake, and I will publish a fulsome apology in this newsletter. If the results show that she is wrong I will donate her stake to HealthWatch [a medical campaigning group], and suggest that she should tell the 1,500 patients on her waiting list that further research has shown that the claimed benefits of her diet have not been observed under controlled conditions. We scientists have a noble tradition of formally withdrawing our publications if subsequent research shows the results are not reproducible—don't we?

Sadly, McKeith—who, to the best of my knowledge, despite all her claims about her extensive "research," has never published in a proper "Pubmed-listed" peer-reviewed academic journal—did not take up this offer to collaborate on a piece of research with a professor of nutrition. Instead Garrow received a call from McKeith's lawyer husband, Howard Magaziner, accusing him of defamation and promising legal action. Garrow, an immensely affable and relaxed old academic, shrugged this off with style. He told me, "I said, 'Sue me.' I'm still waiting." His offer of one thousand pounds still stands.

But there is one vital issue we have not yet covered. Because despite the way McKeith seems to respond to criticism or questioning of her ideas, the unusually complicated story of her qualifications, despite her theatrical abusiveness, and the public humiliation pantomime of her shows, in which the emotionally vulnerable and obese cry on television, despite her apparently misunderstanding some of the most basic aspects of high school biology, despite doling out "scientific" advice in a white coat, despite the dubious quality of the work she presents as somehow being of "academic" standard, despite the unpleasantness of the food she endorses, there are still many who will claim: "You can say what you like about McKeith, but she has improved the people's diet."

On this, let me be very clear, for I will say it once again: anyone who tells you to eat more fresh fruits and vegetables is all right by me. If that were the end of it, I'd be nutritionists' biggest fan, because I'm all in favor of "evidence-based interventions to improve the nation's health," as they used to say to us in medical school.

Let's look at the evidence. Diet has been studied very extensively, and there are some things that we know with a fair degree of certainty: there is reasonably convincing evidence that having a diet rich in fresh fruit and vegetables, with natural sources of dietary fiber, avoiding obesity, moderating one's intake of alcohol, cutting out cigarettes, and taking physical exercise are protective against such things as cancer and heart disease.

Nutritionists don't stop there, because they can't; they have to manufacture complication, to justify the existence of their profession. These new nutritionists have a major commercial problem with the evidence. There's nothing very professional or proprietary about "Eat your greens," so they have had to push things further. But unfortunately for them, the technical, confusing, overcomplicated, tinkering interventions that they promote—the enzymes, the exotic berries—are very frequently not supported by convincing evidence.

That's not for lack of looking. This is not a case of the medical hegemony's neglecting to address the holistic needs of the people. In many cases the research has been done and has shown that the more specific claims of nutritionists are actually wrong. The fairy tale of antioxidants is a perfect example. Sensible dietary practices, which we all know about, still stand. But the unjustified, unnecessary overcomplication of this basic dietary advice is, to my mind, one of the greatest crimes of the nutritionist movement. As I have said, I don't think it's excessive to talk about consumers paralyzed with confusion in supermarkets.

But what can you do? There's the rub. The most important take-home message with diet and health is that anyone who ever expresses anything with certainty is basically wrong, because the evidence for cause and effect in this area is almost always weak and circumstantial, and changing an individual person's diet may not even be where the action is.

What is the best evidence on the benefits of changing an individual person's diet? There have been randomized controlled trials, for example, in which you take a large group of people, change their diets, and compare their health outcomes with another group, but these have generally shown very disappointing results.

The Multiple Risk Factor Intervention Trial was one of the largest medical research projects ever undertaken in the history of mankind, involving over 12,866 men at risk of cardiovascular events, who went through the trial over seven years. These people were subjected to a phenomenal palaver: questionnaires, twenty-four-hour dietary recall interviews, three-day food records, regular visits, and more. On top of this, there were hugely energetic interventions that were supposed to change the lives of individuals, but which by necessity required that whole families' eating patterns were transformed: so there were weekly group information sessions for participants—and their wives—individual work, counseling, an intensive education program, and more. The results, to everyone's disappointment, showed no benefit over the control group (who were not told to change their diet). The Women's Health Initiative was another huge randomized controlled trial into dietary change, and it gave similarly gave negative results. They all tend to.

Why should this be? The reasons are fascinating, and a window into the complexities of changing health behavior. I can discuss only a few here, but if you are genuinely interested in preventive medicine—and you can cope with uncertainty and the absence of quick-fix gimmicks—then may I recommend you pursue a career in it, because you won't get on television, but you will be both dealing in sense and doing good.

The most important thing to notice is that these trials require people to turn their entire lives upside down and for about a decade. That's a big ask; it's hard enough to get people signed up for participating in a seven-week trial, let alone one that lasts seven years, and this has two interesting effects. First, your participants probably won't change their diets as much as you want them to, but far from being a failing, this is actually an excellent illustration of what happens in the real world: individual people do not, in reality, change their diets at the drop of a hat, alone, as individuals, for the long term. A dietary change probably requires a change in lifestyle, shopping habits, maybe even what's in the shops, how you use your time; it might even require that you buy some cooking equipment, change how your family relates to one another, change your work style, and so on.

Second, the people in your "control group" will change their diets too; remember, they've agreed voluntarily to take part in a hugely intrusive seven-year-long project that could require massive lifestyle changes, so they may have a greater interest in health than the rest of your population. More than that, they're also being weighed, measured, and quizzed about their diet, all at regular

intervals. Diet and health are suddenly much more at the forefront of their minds. They will change too.

This is not to rubbish the role of diet in health—I bend over
backward to find some good in these studies—but it does reflect
one of the most important issues, which is that you might not start
with goji berries, or vitamin pills, or magic enzyme powders, and in
fact, you might not even start with an individual's changing his or
her diet. Piecemeal individual life changes, which go against the
grain of your own life and your environment, are hard to make and
even harder to maintain. It's important to see the individual—and
the dramatic claims of all lifestyle nutritionists, for that matter—in
a wider social context.

Reasonable benefits have been shown in intervention studies—
like the North Karelia Project in Finland—in which the public
health gang have moved themselves in lock, stock, and barrel to set
about changing everything about an entire community's behavior,
liaising with businesses to change the food in shops, modifying whole
lifestyles, employing community educators and advocates, improving health care provision, and more, producing some benefits, if you
accept that the methodology used justifies a causal inference. (It's
tricky to engineer a control group for this kind of study, so you have
to make pragmatic decisions about study design, but read it online
and decide for yourself: I'd call it a large and promising case study.)

There are fairly good grounds to believe that many of these
lifestyle issues are in fact better addressed at the societal level.
One of the most significant "lifestyle" causes of death and disease,
after all, is social class. To take a concrete example, in the Bronx
of New York City, a poor multiracial borough where the average
salary is around $35,000, 25 percent of the population is obese and
27 percent have serious health problems. Just across the East River
in Manhattan, where the billionaire Michael Bloomberg lives, surrounded by other wealthy and middle-class people, just 15 percent
are obese and 20 percent have serious health problems.

The reason for this phenomenal disparity in health is not that the people in Manhattan are careful to eat goji berries and a handful of Brazil nuts every day, thus ensuring they're not deficient in selenium, as per nutritionists' advice. That's a fantasy and in some respects one of the most destructive features of the whole nutritionist project; it's a distraction from the real causes of ill health, but also—do stop me if I'm pushing this too far—in some respects, a manifesto of right-wing individualism. You are what you eat, and people die young because they deserve it. *They* choose death, through ignorance and laziness, but *you* choose life, fresh fish, olive oil, and that's why you're healthy. You're going to see eighty. You deserve it. Not like *them*.

Back in the real world, genuine public health interventions to address the social and lifestyle causes of disease are far less lucrative, and far less of a spectacle, than anything a vitamin pill peddler, or a nutritionist, would care to engage with. Who puts the issue of social inequality driving health inequality onto our screens? Where's the human interest in prohibiting the promotion of bad foods, facilitating access to healthier foods by means of taxation or maintaining a clear labeling system?

Where is the spectacle in "enabling environments" that naturally promote exercise, or urban planning that prioritizes cyclists, pedestrians, and public transport over the car? Or in reducing the ever-increasing inequality between senior executive and shop floor pay? When did you ever hear about elegant ideas like walking school buses, or were stories about their benefits crowded out by the latest urgent front-page food fad news?

I don't expect nutritionist, or pill peddlers, or anyone in the media to address a single one of these issues, and if you're honest, neither do you.

8

THE DOCTOR WILL SUE YOU NOW

This chapter did not appear in the original British edition of this book, because for fifteen months leading up to September 2008 the vitamin pill entrepreneur Matthias Rath was suing me personally, and *The Guardian*, for libel. This strategy brought only mixed success. For all that nutritionists may fantasize in public that any critic is somehow a pawn of big pharma, in private they would do well to remember that like many my age who work in the public sector, I don't own an apartment. *The Guardian* generously paid for the lawyers, and in September 2008 Rath dropped his case, which had cost in excess of $770,000 to defend. He eventually paid $365,000, leaving *The Guardian* with a large shortfall. Nobody will ever repay me for the endless meetings, the time off work, or the days spent poring over tables filled with endlessly cross-referenced court documents.

On this last point there is, however, one small consolation, and I will spell it out as a cautionary tale: I now know more about Matthias Rath than almost any other person alive. My notes, references, and witness statements, boxed up in the room where I am sitting right now, make a pile as tall as the man himself, and what

I will write here is only a tiny fraction of the fuller story that is waiting to be told about him. This chapter, I should also mention, is available free online for anyone who wishes to see it.

Matthias Rath takes us rudely outside the contained, almost academic distance of this book. For the most part we've been interested in the intellectual and cultural consequences of bad science, the made-up facts in national newspapers, dubious academic practices in universities, some foolish pill peddling, and so on. But what happens if we take these sleights of hand, these pill-marketing techniques, and transplant them out of our decadent Western context into a situation where things really matter?

In an ideal world this would be only a thought experiment.

AIDS is the opposite of anecdote. Already 25 million people have died from it, 3 million in the last year alone, and 500,000 of those deaths were children. In South Africa it kills 300,000 people every year: that's 800 people every day, or 1 every two minutes. This one country has 6.3 million people who are HIV positive, including 30 percent of all pregnant women. There are 1.2 million AIDS orphans under the age of seventeen. Most chillingly of all, this disaster has appeared suddenly, and while we were watching: in 1990, just 1 percent of adults in South Africa were HIV positive. Just ten years later, the figure had risen to 25 percent.

It's hard to mount an emotional response to raw numbers, but on one thing I think we would agree: if you were to walk into a situation with that much death, misery, and disease, you would be very careful to make sure that you knew what you were talking about. For the reasons you are about to read, I suspect that Matthias Rath missed the mark.

This man, we should be clear, is our responsibility. Born and raised in Germany, Rath was the head of cardiovascular research at the Linus Pauling Institute in Menlo Park, California, and even then he had a tendency toward grand gestures, publishing a paper in *The Journal of Orthomolecular Medicine* in 1992 titled "A Unified Theory of Human Cardiovascular Disease Leading the Way to the Abolition of This Disease as a Cause for Human Mortality." The unified theory was high-dose vitamins.

He first developed a power base from sales in Europe, selling his pills with tactics that will be very familiar to you from the rest of this book, albeit slightly more aggressive. In the U.K., his ads claimed that "90 percent of patients receiving chemotherapy for cancer die within months of starting treatment" and suggested that three million lives could be saved if cancer patients stopped being treated by conventional medicine. The pharmaceutical industry was deliberately letting people die for financial gain, he explained. Cancer treatments were "poisonous compounds" with "not even one effective treatment."

The decision to embark on treatment for cancer can be the most difficult that an individual or a family will ever take, representing a close balance between well-documented benefits and equally well-documented side effects. Ads like these might play especially strongly on your conscience if your mother had just lost all her hair to chemotherapy, for example, in the hope of staying alive just long enough to see your son speak.

There was some limited regulatory response in Europe, but it was generally as weak as that faced by the other characters in this book. The Advertising Standards Authority criticized one of his ads in the U.K., but that is essentially all it is able to do. Rath was ordered by a Berlin court to stop claiming that his vitamins could cure cancer or face a $335,000 fine.

But sales were strong, and Matthias Rath still has many supporters in Europe, as you will shortly see. He walked into South Africa with all the acclaim, self-confidence, and wealth he had amassed as a successful vitamin pill entrepreneur in Europe and America, and began to take out full-page ads in newspapers.

"The answer to the AIDS epidemic is here," he proclaimed. Antiretroviral drugs were poisonous and a conspiracy to kill pa-

tients and make money. STOP AIDS GENOCIDE BY THE DRUGS CAR-
TEL, said one headline. "Why should South Africans continue to
be poisoned with AZT? There is a natural answer to AIDS." The
answer came in the form of vitamin pills. "Multivitamin treatment
is more effective than any toxic AIDS drug." "Multivitamins cut the
risk of developing AIDS in half."

Rath's company ran clinics reflecting these ideas, and in 2005
he decided to run a trial of his vitamins in a township near Cape
Town called Khayelitsha, giving his own formulation, VitaCell, to
people with advanced AIDS. In 2008 this trial was declared illegal by the Cape High Court of South Africa. Although Rath says
that none of his participants had been on antiretroviral drugs, some
relatives have given statements saying that they had been and
were actively told to stop using them.

Tragically, Matthias Rath had taken these ideas to exactly the
right place. Thabo Mbeki, the president of South Africa at the
time, was well known as an AIDS dissident, and to international
horror, while people died at the rate of one every two minutes in
his country, he gave credence and support to the claims of a small
band of campaigners who variously claim that AIDS does not exist, that it is not caused by HIV, that antiretroviral medication
does more harm than good, and so on.

At various times during the peak of the AIDS epidemic in
South Africa its government argued that HIV is not the cause of
AIDS and that antiretroviral drugs are not useful for patients. It
refused to roll out proper treatment programs, it refused to accept
free donations of drugs, and it refused to accept grant money from
the Global Fund to buy drugs.

One study estimates that if the South African national government had used antiretroviral drugs for prevention and treatment at the same rate as the Western Cape Province (which defied
national policy on the issue), around 171,000 new HIV infections
and 343,000 deaths could have been prevented between 1999 and

2007. Another study estimates that between 2000 and 2005 there were 330,000 unnecessary deaths, 2.2 million person-years lost, and 35,000 babies unnecessarily born with HIV because of the failure to implement a cheap and simple mother-to-child transmission prevention program. Between one and three doses of an ARV drug can reduce transmission dramatically. The cost is negligible. It was not available.

Interestingly, Matthias Rath's colleague and employee, a South African barrister named Anthony Brink, takes the credit for introducing Thabo Mbeki to many of these ideas. Brink stumbled on the AIDS dissident material in the mid-1990s and, after much surfing and reading, became convinced that it must be right. In 1999 he wrote an article about AZT in a Johannesburg newspaper titled "A Medicine from Hell." This led to a public exchange with a leading virologist. Brink contacted Mbeki, sending him copies of the debate, and was welcomed as an expert. This is a chilling testament to the danger of elevating cranks by engaging with them.

In his initial letter of motivation for employment to Matthias Rath, Brink described himself as "South Africa's leading AIDS dissident, best known for my whistle-blowing exposé of the toxicity and inefficacy of AIDS drugs, and for my political activism in this regard, which caused President Mbeki and Health Minister Dr. Tshabalala-Msimang to repudiate the drugs in 1999."

In 2000, the now-infamous International AIDS Conference took place in Durban. Mbeki's presidential advisory panel beforehand was packed with AIDS dissidents, including Peter Duesberg and David Rasnick. On the first day, Rasnick suggested that all HIV testing should be banned on principle and that South Africa should stop screening supplies of blood for HIV. "If I had the power to outlaw the HIV antibody test," he said, "I would do it across the board." When African physicians gave testimony about the drastic change AIDS had caused in their clinics and hospitals, Rasnick said he had not seen "any evidence" of an AIDS catastrophe. The media were

not allowed in, but one reporter from *The Village Voice* was present. Peter Duesberg, he said, "gave a presentation so removed from
African medical reality that it left several local doctors shaking
their heads." It wasn't AIDS that was killing babies and children,
said the dissidents; it was the antiretroviral medication.

President Mbeki sent a letter to world leaders comparing the
struggle of the "AIDS dissidents" with the struggle against apartheid. *The Washington Post* described the reaction at the White
House: "So stunned were some officials by the letter's tone and
timing—during final preparations for July's conference in Durban—
that at least two of them, according to diplomatic sources, felt
obliged to check whether it was genuine." Hundreds of delegates
walked out of Mbeki's address to the conference in disgust, but many
more described themselves as dazed and confused. More than five
thousand researchers and activists around the world signed up to
the Durban Declaration, a document that specifically addressed
and repudiated the claims and concerns—at least the more moderate ones—of the AIDS dissidents. Specifically, it addressed the
charge that people were simply dying of poverty:

> The evidence that AIDS is caused by HIV-1 or HIV-2 is
> clear-cut, exhaustive and unambiguous . . . As with any
> other chronic infection, various co-factors play a role in
> determining the risk of disease. Persons who are malnour-
> ished, who already suffer other infections or who are older,
> tend to be more susceptible to the rapid development of
> AIDS following HIV infection. However, none of these
> factors weaken the scientific evidence that HIV is the
> sole cause of AIDS . . . Mother-to-child transmission can
> be reduced by half or more by short courses of antiviral
> drugs . . . What works best in one country may not be ap-
> propriate in another. But to tackle the disease, everyone
> must first understand that HIV is the enemy. Research,

not myths, will lead to the development of more effective
and cheaper treatments.

It did them no good. Until 2003 the South African government
refused, as a matter of principle, to roll out proper antiretroviral
medication programs, and even then the process was halfhearted.
This madness was overturned only after a massive campaign by
grassroots organizations such as the Treatment Action Campaign,
but even after the ANC cabinet voted to allow medication to be
given, there was still resistance. In mid-2005, at least 85 percent of
HIV positive people who needed antiretroviral drugs were still refused them. That's around a million people.

This resistance, of course, went deeper than just one man;
much of it came from Mbeki's health minister, Manto Tshabalala-
Msimang. An ardent critic of medical drugs for HIV, she would
cheerfully go on television to talk up their dangers, talk down their
benefits, and became irritable and evasive when asked how many
patients were receiving effective treatment. She declared in 2005
that she would not be "pressured" into meeting the target of three
million patients on antiretroviral medication, that people had ignored the importance of nutrition, and that she would continue to
warn patients of the side effects of antiretrovirals, saying: "We have
been vindicated in this regard. We are what we eat."

It's an eerily familiar catchphrase. Tshabalala-Msimang also
went on record to praise the work of Matthias Rath and refused
to investigate his activities. Most joyfully of all, she was a staunch
advocate of the kind of weekend glossy magazine–style nutritionism that will by now be very familiar to you.

The remedies she advocated for AIDS are beetroot, garlic,
lemons, and African potatoes. A fairly typical quote, from the
health minister in a country where eight hundred people die every
day from AIDS, was this: "Raw garlic and a skin of the lemon—
not only do they give you a beautiful face and skin but they also

protect you from disease." South Africa's stand at the 2006 World AIDS Conference in Toronto was described by delegates as the "salad stall." It consisted of some garlic, some beetroot, African potato, and assorted other vegetables. Some boxes of antiretroviral drugs were added later, but they were reportedly borrowed at the last minute from other conference delegates.

Alternative therapists like to suggest that their treatments and ideas have not been sufficiently researched. As you now know, this is often untrue, and in the case of the health minister's favored vegetables, research had indeed been done, with results that were far from promising. Interviewed on SABC about this, Tshabalala-Msimang gave the kind of responses you'd expect to hear at any North London dinner party discussion of alternative therapies.

First she was asked about work from the University of Stellenbosch that suggested that her chosen plant, African potato, might be actively dangerous for people on AIDS drugs. One study on African potato in HIV had to be terminated prematurely, because the patients who received the plant extract developed severe bone marrow suppression and a drop in their CD4 cell count—which is a bad thing—after eight weeks. On top of this, when extract from the same vegetable was given to cats with feline immunodeficiency virus, they succumbed to full-blown feline AIDS faster than their nontreated controls. African potato does not look like a good bet.

Tshabalala-Msimang disagreed; the researchers should go back to the drawing board, and "investigate properly." Why? Because HIV positive people who used African potato had shown improvement, and they had said so themselves. If a person says he or she is feeling better, should this be disputed, she demanded to know, merely because it had not been proved scientifically? "When a person says she or he is feeling better, I must say 'No, I don't think you are feeling better'? 'I must rather go and do science on you'?" Asked whether there should be a scientific basis to her views, she replied: "Whose science?"

And there, perhaps, is a clue, if not exoneration. This is a continent that has been brutally exploited by the developed world, first by empire and then by globalized capital. Conspiracy theories about AIDS and Western medicine are not entirely absurd in this context. The pharmaceutical industry has indeed been caught performing drug trials in Africa that would be impossible anywhere in the developed world. Many find it suspicious that black Africans seem to be the biggest victims of AIDS and point to the biological warfare programs set up by the apartheid governments; there have also been suspicions that the scientific discourse of HIV/AIDS might be a device, a Trojan horse for spreading even more exploitative Western political and economic agendas around a problem that is simply one of poverty.

And these are new countries, for which independence and self-rule are recent developments, which are struggling to find their commercial feet and true cultural identity after centuries of colonization. Traditional medicine represents an important link with an autonomous past, besides which, antiretroviral medications have been unnecessarily—offensively, absurdly—expensive, and until moves to challenge this became partially successful, many Africans were effectively denied access to medical treatment as a result.

It's very easy for us to feel smug and to forget that we all have our own strange cultural idiosyncrasies that prevent us from taking up sensible public health programs. For examples, we don't even have to look as far as MMR. There is a good evidence base to show that needle exchange programs reduce the spread of HIV, but this strategy has been rejected time and again in favor of "Just say no." Development charities funded by U.S. Christian groups refuse to engage with birth control, and any suggestion of abortion, even in countries where being in control of your own fertility could mean the difference between success and failure in life, is met with a cold, pious stare. These impractical moral principles are so deeply entrenched that PEPFAR, the U.S. Presidential Emergency Plan for

AIDS Relief, has insisted that every recipient of international aid money must sign a declaration expressly promising not to have any involvement with sex workers.

We mustn't appear insensitive to the Christian value system, but it seems to me that engaging sex workers is almost the corner-stone of any effective AIDS policy. Commercial sex is frequently the "vector of transmission," and sex workers are a very high-risk population; but there are also more subtle issues at stake. If you secure the legal rights of prostitutes to be free from violence and discrimination, you empower them to demand universal condom use, and that way you can prevent HIV from being spread into the whole community. This is where science meets culture. But perhaps even to your own friends and neighbors, in whatever suburban idyll has become your home, the moral principle of abstinence from sex and drugs is more important than people dying of AIDS, and per-haps, then, they are no less irrational than Thabo Mbeki.

So this was the situation into which the vitamin pill entrepre-neur Matthias Rath inserted himself, prominently and expen-sively, with the wealth he had amassed from Europe and America, exploiting anticolonial anxieties with no sense of irony, although he was a white man offering pills made in a factory abroad. His ads and clinics were a tremendous success. He began to tout individ-ual patients as evidence of the benefits that could come from vita-min pills, although in reality some of his most famous success stories have died of AIDS. When asked about the deaths of Rath's star patients, Health Minister Tshabalala-Msimang replied: "It doesn't necessarily mean that if I am taking antibiotics and I die, that I died of antibiotics."

She was not alone: South Africa's politicians have consistently refused to step in, Rath claims the support of the government, and its most senior figures have refused to distance themselves from his operations or to criticize his activities. Tshabalala-Msimang went on the record to state that the Rath Foundation "are not undermining the government's position. If anything, they are support-ing it."

In 2005, exasperated by government inaction, a group of 199 leading medical practitioners in South Africa signed an open let-ter to the health authorities of the Western Cape, pleading for action on the Rath Foundation. "Our patients are being inundated with propaganda encouraging them to stop life-saving medicine," it said. "Many of us have had experiences with HIV-infected pa-tients who have had their health compromised by stopping their antiretrovirals due to the activities of this Foundation."

Rath's ads continue unabated. He even claimed that his ac-tivities were endorsed by huge lists of sponsors and affiliates, in-cluding the World Health Organization, UNICEF, and UNAIDS. All have issued statements flatly denouncing his claims and acti-vities. The man certainly has chutzpah.

. His ads are also rich with detailed scientific claims. It would be wrong of us to neglect the science in this story, so we should follow some through, specifically those which focused on a Har-vard study in Tanzania. He described this research in full-page advertisements, some of which have appeared in *The New York Times* and the *Herald Tribune*. He refers to these paid ads, I should mention, as if he had received flattering news coverage in the same papers. Anyway, this research showed that multivitamin supple-ments can be beneficial in a developing world population with AIDS. There's no problem with that result, and there are plenty of reasons to think that vitamins might have some benefit for a sick and frequently malnourished population.

The researchers enrolled 1,078 HIV positive pregnant women and randomly assigned them to have either a vitamin supplement or a placebo. Notice once again, if you will, that this is another large, well-conducted, publicly funded trial of vitamins, conducted by mainstream scientists, contrary to the claims of nutritionists that such studies do not exist.

The women were followed up for several years, and at the end of the study, 25 percent of those on vitamins were severely ill or dead, compared with 31 percent of those on placebo. There was also a statistically significant benefit in CD4 cell count (a measure of HIV activity) and viral loads. These results were in no sense dramatic—and they cannot be compared with the demonstrable lifesaving benefits of antiretrovirals—but they did show that improved diet or cheap generic vitamin pills could represent a simple and relatively inexpensive way to marginally delay the need to start HIV medication in some patients.

In the hands of Rath, this study became evidence that vitamin pills are superior to medication in the treatment of HIV/AIDS, that antiretroviral therapies "severely damage all cells in the body—including white blood cells," and worse, that they were "thereby not improving but rather worsening immune deficiencies and expanding the AIDS epidemic." The researchers from the Harvard School of Public Health were so horrified that they put together a press release setting out their support for medication and stating starkly, with unambiguous clarity, that Matthias Rath had misrepresented their findings. Media regulators failed to act.

To outsiders the story is baffling and terrifying. The United Nations has condemned Rath's ads as "wrong and misleading." "This guy is killing people by luring them with unrecognized treatment without any scientific evidence," said Eric Goemaere, head of Médecins sans Frontières SA, a man who pioneered antiretroviral therapy in South Africa. Rath sued him.

It's not just MSF that Rath has gone after. He has also brought time-consuming, expensive, stalled, or failed cases against a professor of AIDS research, critics in the media, and others.

His most heinous campaign has been against the Treatment Action Campaign. For many years this has been the key organization campaigning for access to antiretroviral medication in South Africa, and it has been fighting a war on four fronts. First, it campaigns against its own government, trying to compel it to roll out treatment programs for the population. Second, it fights against the pharmaceutical industry, which claims that it needs to charge full price for its products in developing countries in order to pay for research and development of new drugs, although, as we shall see, out of its $550 billion global annual revenue, the pharmaceutical industry spends twice as much on promotion and administration as it does on research and development. Third, it is a grassroots organization, made up largely of black women from townships who do important prevention and treatment literacy work on the ground, ensuring that people know what is available and how to protect themselves. Lastly, it fights against people who promote the type of information peddled by Matthias Rath and his like.

Rath has taken it upon himself to launch a massive campaign against this group. He distributes advertising material against them, saying, "Treatment Action Campaign medicines are killing you" and "Stop AIDS genocide by the drug cartel," claiming—as you will guess by now—that there is an international conspiracy by pharmaceutical companies intent on prolonging the AIDS crisis in the interests of their own profits by giving medication that makes people worse. TAC must be a part of this, goes the reasoning, because it criticizes Matthias Rath. Just like me writing on Gillian McKeith, TAC is perfectly in favor of good diet and nutrition. But in Rath's promotional literature it is a front for the pharmaceutical industry, a "Trojan horse" and a "running dog." TAC has made a full disclosure of its funding and activities, showing no such connection. Rath has presented no evidence to the contrary, and has even lost a court case over the issue, but will not let it lie. In fact, he presents the loss of this court case as if it were a victory.

The founder of TAC is a man called Zackie Achmat, and he is the closest thing I have to a hero. He is South African, and Coloured, by the nomenclature of the apartheid system in which he grew up. At the age of fourteen he tried to burn down his school, and you might have done the same in similar circumstances. He has been arrested and imprisoned under South Africa's violent, brutal white regime, with all that entailed. He is also gay and HIV positive, and he refused to take antiretroviral medication until it was widely available to all on the public health system, even when he was dying of AIDS, even when he was personally implored to save himself by Nelson Mandela, a public supporter of antiretroviral medication and Achmat's work.

And now, at last, we come to the lowest point of this whole story, not merely for Matthias Rath's movement but for the alternative therapy movement around the world as a whole. In 2007, with a huge public flourish, to great media coverage, Rath's former employee Anthony Brink filed a formal complaint against Zackie Achmat, the head of TAC. Bizarrely, he filed this complaint with the International Criminal Court at The Hague, accusing Achmat of genocide for successfully campaigning to get access to HIV drugs for the people of South Africa.

It's hard to explain just how influential the AIDS dissidents are in South Africa. Brink is a barrister, a man with important friends, and his accusations were reported in the national news media—and in some corners of the Western gay press—as a serious news story. I do not believe that any one of those journalists who reported on it can possibly have read Brink's indictment to the end. I have. The first fifty-seven pages present familiar anti-medication and AIDS dissident material. But then, on page fifty-eight, this "indictment" document suddenly deteriorates into something altogether more vicious and unhinged, as Brink sets out what he believes would be an appropriate punishment for Zackie. Because I do not wish to be accused of selective editing, I will now reproduce for you that entire section, unedited, so you can see and feel it for yourself.

Appropriate Criminal Sanction

In view of the scale and gravity of Achmat's crime and his direct personal criminal culpability for "the deaths of thousands of people," to quote his own words, it is respectfully submitted that the International Criminal Court ought to impose on him the highest sentence provided by Article 77.1(b) of the Rome Statute, namely to permanent confinement in a small white steel and concrete cage, bright fluorescent light on all the time to keep an eye on him, his warders putting him out only to work every day in the prison garden to cultivate nutrient-rich vegetables, including when it's raining. In order for him to repay his debt to society, with the ARVs he claims to take administered daily under close medical watch at the full prescribed dose, morning noon and night, without interruption, to prevent him faking that he's being treatment compliant, pushed if necessary down his forced-open gullet with a finger, or, if he bites, kicks and screams too much, dripped into his arm after he's been restrained on a gurney with cable ties around his ankles, wrists and neck, until he gives up the ghost on them, so as to eradicate this foulest, most loathsome, unscrupulous and malevolent blight on the human race, who has plagued and poisoned the people of South Africa, mostly black, mostly poor, for nearly a decade now, since the day he and his TAC first hit the scene.

Signed at Cape Town, South Africa, on 1 January 2007

Anthony Brink

The document was described by the Rath Foundation as "entirely valid and long overdue."

This story isn't about Matthias Rath, or Anthony Brink, or Zackie Achmat, or even South Africa. It is about the culture of how ideas work, and how that can break down. Doctors criticize other doctors; academics criticize academics; politicians criticize politicians. That's normal and healthy; it's how ideas improve. Matthias Rath is an alternative therapist, made in Europe. He is every bit the same as some of the operators that we have seen in this book. He is from their world.

Despite the extremes of this case, not one single alternative therapist or nutritionist, anywhere in the world, has stood up to criticize any single aspect of the activities of Matthias Rath and his colleagues. In fact, far from it; he continues to be feted to this day. I have sat in true astonishment and watched leading figures of the U.K.'s alternative therapy movement applaud Matthias Rath at a public lecture (I have it on video, just in case there's any doubt). Homeopaths' mailouts continue to promote his work. Not one person will step forward and dissent.

The alternative therapy movement as a whole has demonstrated itself to be so dangerously, systemically incapable of critical self-appraisal that it cannot step up even in a case like that of Rath; in that count I include tens of thousands of practitioners, writers, administrators, and more. This is how ideas go badly wrong. In the conclusion to this book, written before I was able to include this chapter, I will argue that the biggest dangers posed by the material we have covered are cultural and intellectual.

I may be mistaken.

9

IS MAINSTREAM MEDICINE EVIL?

So that was the alternative therapy industry. Its practitioners' claims are made directly to the public, so they have greater cultural currency, and while they use the same tricks of the trade as the pharmaceutical industry, as we have seen along the way, their strategies and errors are more transparent, so they make for a neat teaching tool. Now, once again, we should raise our game.

For this chapter you will also have to rise above your own narcissism. We will not be talking about the fact that your doctor is sometimes rushed or rude to you. We will not be talking about the fact that nobody could work out what was wrong with your knee, and we will not even be discussing the time that someone misdiagnosed your grandfather's cancer, and he suffered unnecessarily for months before a painful, bloody, undeserved, and undignified death at the end of a productive and loving life.

Terrible things happen in medicine, when it goes right as well as when it goes wrong. Everybody agrees that we should work to minimize the errors, everybody agrees that doctors are sometimes terrible; if the subject fascinates you, then I encourage you to buy one of the libraries' worth of books on clinical governance. Doc-

tors can be awful, and mistakes can be murderous, but the philosophy driving evidence based medicine is not. How well does it
work? One thing you could measure is how much medical practice
is evidence based. This is not easy. From the state of current
knowledge, around 13 percent of all *treatments* have good evidence, and a further 21 percent are likely to be beneficial. This
sounds low, but it seems the more common treatments tend to
have a better evidence base. Another way of measuring is to look
at how much medical *activity* is evidence based, taking consecutive patients, in a hospital outpatients' clinic, for example, looking
at their diagnosis, what treatments they were given, and then
looking at whether those treatment decisions were based on evidence. These real-world studies give a more meaningful figure: lots
were done in the 1990s, and it turns out, depending on specialty,
that between 50 and 80 percent of all medical activity is "evidence
based." It's still not great, and if you have any ideas on how to improve that, do please write about it. Another good measure is what
happens when things go wrong. The *British Medical Journal*, to
take an example, is one of the biggest medical journals in the
world. It recently announced the three most popular papers from
its archive for 2005, according to an audit that assessed their use
by readers, the number of times they were referenced by other academic papers, and so on. Each of these papers had a criticism of a
drug, a drug company, or a medical activity as its central theme.

We can go through them briefly, so you can see for yourself
how relevant the biggest papers from the most important medical
journal are to your needs. The top-scoring paper was a case-control
study that showed that patients had a higher risk of heart attack if
they were taking the drugs rofecoxib (Vioxx), diclofenac, or ibuprofen. At number two was a large meta-analysis of drug company
data, which showed no evidence that SSRI antidepressants increase
the risk of suicide, but found weak evidence for an increased risk
of deliberate self-harm. In third place was a systematic review that

showed an *association* between suicide attempts and the use of SSRIs, and critically highlighted some of the inadequacies around the reporting of suicides in clinical trials.

This is critical self-appraisal, and it is very healthy, but you will notice something else: all those studies revolve around situations in which drug companies withheld or distorted evidence. How does this happen?

THE PHARMACEUTICAL INDUSTRY

The tricks of the trade that we'll discuss in this chapter are probably more complicated than most of the other stuff in the book, because we will be making technical critiques of an industry's professional literature. There is also, of course, a whole separate book to be written about the marketing techniques. In the United States and New Zealand (but nowhere else in the developed world) drug companies are allowed to advertise their pills directly to the public, often with bizarre results, including my own personal favorite, "Clomicalm tablets are the only medication approved for the treatment of separation anxiety in dogs." The U.S. pharmaceutical industry's annual spend on promotion is more than three billion dollars, and it works, increasing prescriptions and doctor visits.

But here we are pulling apart not the marketing but the distortions in the science, and the tricks they play on doctors, who are harder to bluff, as an audience. This means that we'll first have to explain some background about how a drug comes to market. This is stuff that you will be taught at school when I become president of the one world government.

Understanding this process is important for one very clear reason: it seems to me that a lot of the stranger ideas people have about medicine derive from an emotional struggle with the very notion of a pharmaceutical industry. Whatever our political leanings, we all feel nervous about profit taking any role in the caring professions, but that feeling has nowhere to go. Big pharma is evil; I would agree with that premise. But because people don't understand exactly *how* big pharma is evil, their anger gets diverted away from valid criticisms—its role in distorting data, for example, or withholding lifesaving AIDS drugs from the developing world—and channeled into infantile fantasies. "Big pharma is evil," goes the line of reasoning; "therefore homeopathy works and the MMR vaccine causes autism." This is probably not helpful.

In the United States, the pharmaceutical industry has been one of the most profitable industries over the last twenty-five years. It only lost its first-place standing in 2003, and is currently in third place after Internet and communications companies. The country spent $227.5 billion a year on pharmaceutical drugs in 2009, and much of that goes on patented drugs, medicines that were released in the last ten years. Globally, the industry is worth more than $800 billion.

People come in many flavors, but all corporations have a duty to maximize their profits, and this often sits uncomfortably with the notion of caring for people. An extreme example comes with AIDS. As I mentioned in passing, drug companies explain that they cannot give AIDS drugs off license to developing world countries, because they need the money from sales for research and development. And yet, of the biggest U.S. companies' $200 billion sales, only 14 percent is spent on R & D, compared with 31 percent on marketing and administration.

The companies also set their prices in ways you might judge to be exploitative. Once your drug comes out, you have around ten years "on patent," as the only person who is allowed to make it. Loratadine, produced by Schering-Plough, is an effective antihistamine drug that does not cause the unpleasant antihistamine side effect of drowsiness. It was a unique treatment for a while and highly in demand. Before the patent ran out, the price of the drug was raised thirteen times in just five years, increasing by over 50 percent. Some might regard this as profiteering.

But the pharmaceutical industry is also currently in trouble. The golden age of medicine has creaked to a halt, as we have said, and the number of new drugs, or "new molecular entities," being registered has dwindled from fifty a year in the 1990s to about twenty now. At the same time, the number of me-too drugs has risen, making up to half of all new drugs.

Me-too drugs are an inevitable function of the market; they are rough copies of drugs that already exist, made by another company, but are different enough for a manufacturer to be able to claim its own patent. They take huge effort to produce and need to be tested (on human participants, with all the attendant risks) and trialed and refined and marketed just like a new drug. Sometimes they offer modest benefits (a more convenient dosing regime, for example), but for all the hard work they involve, they don't generally represent a significant breakthrough in human health. They are merely a breakthrough in making money. Where do all these drugs come from?

THE JOURNEY OF A DRUG

First of all, you need an idea for a drug. This can come from any number of places: a molecule in a plant, a receptor in the body that you think you can build a molecule to interface with, an old drug that you've tinkered with, and so on. This part of the story is extremely interesting, and I recommend doing a degree in it. When you think you have a molecule that might be a runner, you test it in animals, to see if it works for whatever you think it should do (and to see if it kills them, of course).

Then you do Phase I, or "first in man," studies on a small number of brave, healthy young men who need money, first to see if it kills them and also to measure basic things like how fast the drug is excreted from the body (this is the phase that went horribly wrong in the TGN1412 tests in 2006, when several young men were seriously injured). If this works, you move to a Phase II trial, in a couple of hundred people with the relevant illness, as a "proof of concept," to work out the dose and to get an idea if it is effective or not. A *lot* of drugs fail at this point, which is a shame; bringing a drug to market costs around five hundred million dollars in total.

Then you do a Phase III trial, in hundreds or thousands of patients, randomized, blinded, comparing your drug against placebo or a comparable treatment, and collect much more data on efficacy and safety. You might need to do a few of these, and then you can apply for a license to sell your drug. After it goes to market, you should be doing more trials, and other people will probably do trials and other studies on your drug too, and we hope everyone will keep their eyes open for any previously unnoticed side effects, ideally reporting them to their country's drug regulator (anyone can report an adverse event to the FDA's MedWatch system online).

Doctors make their rational decision on whether they want to prescribe a drug on the basis of how good it has been shown to be in trials, how bad the side effects are, and sometimes cost. Ideally they will get their information on efficacy from studies published in peer-reviewed academic journals or from other material like textbooks and review articles which are themselves based on primary research like trials. At worst, they will rely on the lies of drug reps and word of mouth.

But drug trials are expensive, so an astonishing 90 percent of clinical drug trials, and 70 percent of trials reported in major medical journals, are conducted or commissioned by the pharmaceutical industry. A key feature of science is that findings should be replicated, but if only one organization is doing the funding, then this feature is lost.

It is tempting to blame the drug companies—although it seems to me that nations and civic organizations are equally at fault here for not coughing up—but wherever you draw your own moral line, the upshot is that drug companies have a huge influence over what gets researched, how it is researched, how the results are reported, how they are analyzed, and how they are interpreted.

Sometimes whole areas can be orphaned because of a lack of money and corporate interest. Homeopaths and vitamin pill quacks would tell you that their pills are good examples of this phenomenon.(That is a moral affront to the better examples.)There are conditions that affect a small number of people, like Creutzfeldt-Jakob disease and Wilson's disease, but more chilling are the diseases that are neglected because they are found only in the developing world, like Chagas' disease (which threatens a quarter of Latin America) and trypanosomiasis (three hundred thousand cases a year, but in Africa). The Global Forum for Health Research estimates that only 10 percent of the world's health burden receives 90 percent of total biomedical research funding.

Often it is simply information that is missing, rather than some amazing new molecule. Eclampsia, say, is estimated to cause fifty thousand deaths in pregnancy around the world each year, and the best treatment, by a huge margin, is cheap, unpatented magnesium sulfate (high doses intravenously, that is, not some alternative medicine supplement, but also not the expensive anticonvulsants that were used for many decades). Although magnesium had been used to treat eclampsia since 1906, its position as the best treatment was only established a century later in 2002, with the help of the World Health Organization, because there was no commercial interest in the research question; nobody has a patent on magnesium, and the majority of deaths from eclampsia are in the developing world. Millions of women have died of the condition since 1906, and many of those deaths were avoidable.

To an extent these are political and development issues, which we should leave for another day, and I have a promise to pay out on: you want to be able to take the skills you've learned about levels of evidence and distortions of research and understand how the pharmaceutical industry distorts data and pulls the wool over our eyes. How would we go about proving this? Overall, it's true, drug company trials are much more likely to produce a positive outcome for their own drugs. But to leave it there would be weak-minded.

What I'm about to tell you is what I teach medical students and doctors—here and there—in a lecture I rather childishly call drug company bullshit. It is in turn what I was taught at medical school,* and I think the easiest way to understand the issue is to put yourself in the shoes of a big pharma researcher.

You have a pill. It's OK, maybe not that brilliant, but a lot of money is riding on it. You need a positive result, but your audience aren't homeopaths, journalists, or the public; they are doctors and academics, who have been trained in spotting the obvious tricks, like "no blinding," or "inadequate randomization." Your sleights of hand will have to be much more elegant, much more subtle, but every bit as powerful.

What can you do?

Well, first, you could study it in winners. Different people respond differently to drugs: old people on lots of medications are often no-hopers, whereas younger people with just one problem are more likely to show an improvement. So study your drug only in the latter group. This will make your research much less applicable to the actual people that doctors are prescribing for, but you hope won't notice. This is so commonplace it is hardly worth giving an example.

*In this subject, like many medics of my generation, I am indebted to the classic textbook *How to Read a Paper* by Professor Trisha Greenhalgh at the University College London. It should be a bestseller. *Testing Treatments* by Imogen Evans, Hazel Thornton, and Iain Chalmers is also a work of great genius, appropriate for a lay audience and amazingly also free to download online from www.jameslindlibrary.org. For committed readers I recommend *Methodological Errors in Medical Research* by Bjorn Andersen. It's extremely long. The subtitle is *An Incomplete Catalogue.*

Next up, you could compare your drug against a useless control. Many people would argue, for example, that you should *never* compare your drug with placebo, because it proves nothing of clinical value. In the real world, nobody cares if your drug is better than a sugar pill; people care only if it is better than the best currently available treatment. But you've already spent hundreds of millions of dollars bringing your drug to market, so stuff that: do lots of placebo-controlled trials, and make a big fuss about them, because they practically guarantee some positive data. Again, this is universal, because almost all drugs will be compared against placebo at some stage in their lives, and drug reps, the people employed by big pharma to bamboozle doctors, love the unambiguous positivity of the graphs these studies can produce.

Then things get more interesting. If you do have to compare your drug with one produced by a competitor—to save face or because a regulator demands it—you could try a sneaky underhand trick: use an inadequate dose of the competing drug, so that patients on it don't do very well, or give a very high dose of the competing drug, so that patients experience lots of side effects, or give the competing drug in the wrong way (perhaps orally when it should be intravenous, and hope most readers don't notice); or you could increase the dose of the competing drug much too quickly, so that the patients taking it get worse side effects. Your drug will shine by comparison.

You might think no such thing could ever happen. If you follow the references in the back, you will find studies in which patients were given really rather high doses of old-fashioned antipsychotic medication (which made the new-generation drugs look as if they were better in terms of side effects) and studies with doses of SSRI antidepressants, which some might consider unusual, to name just a couple of examples. I know. It's slightly incredible.

Of course, another trick you could pull with side effects is simply not to ask about them, or rather—since you have to be sneaky in this field—you could be careful about how you ask. Here is an example. SSRI antidepressant drugs cause sexual side effects, fairly commonly, including anorgasmia. We should be clear (I'm trying to phrase this as neutrally as possible): I *really* enjoy the sensation of orgasm. It's important to me, and everything I experience in the world tells me that this sensation is important to other people too. Wars have been fought, essentially, for the sensation of orgasm. There are evolutionary psychologists who would try to persuade you that the entirety of human culture and language is driven, in large part, by the pursuit of the sensation of orgasm. Losing it seems like an important side effect to ask about.

And yet various studies have shown that the reported prevalence of anorgasmia in patients taking SSRI drugs varies between 2 percent and 73 percent, depending primarily on how you ask: a casual, open-ended question about side effects, for example, or a careful and detailed inquiry. One three-thousand-subject review on SSRIs simply did not list any sexual side effects on its twenty-three–item side effect table. Twenty-three other things were more important, according to the researchers, than losing the sensation of orgasm. I have read them. They are not.

But back to the main outcomes. And here is a good trick: instead of a real-world outcome, like death or pain, you could always use a "surrogate outcome," which is easier to attain. If your drug is supposed to reduce cholesterol and so prevent cardiac deaths, for example, don't measure cardiac deaths; measure reduced cholesterol instead. That's much easier to achieve than a reduction in cardiac deaths, and the trial will be cheaper and quicker to do, so your result will be cheaper *and* more positive. Result!

Now you've done your trial, and despite your best efforts, things have come out negative. What can you do? Well, if your trial has been good overall, but has thrown out a few negative results, you could try an old trick: don't draw attention to the disappointing data by putting it on a graph. Mention it briefly in the text, and ignore it when drawing your conclusions. (I'm so good at this I scare myself. Comes from reading too many rubbish trials.)

If your results are completely negative, don't publish them at all, or publish them only after a long delay. This is exactly what the drug companies did with the data on SSRI antidepressants: they hid the data suggesting they might be dangerous, and they buried the data showing them to perform no better than placebo. If you're really clever and have money to burn, then after you get disappointing data, you could do some more trials with the same protocol, in the hope that they will be positive; then try to bundle all the data up together, so that your negative data is swallowed up by some mediocre positive results.

Or you could get really serious, and start to manipulate the statistics. For two pages only, this book will now get quite nerdy. I understand if you want to skip it, but know that it is here for the doctors who bought the book to laugh at homeopaths. Here are the classic tricks to play in your statistical analysis to make sure your trial has a positive result.

IGNORE THE PROTOCOL ENTIRELY

Always assume that any correlation *proves* causation. Throw all your data into a spreadsheet program, and report—as significant— any relationship between anything and everything if it helps your case. If you measure enough, some things are bound to be positive just by sheer luck.

PLAY WITH THE BASELINE

Sometimes, when you start a trial, quite by chance the treatment group is already doing better than the placebo group. If so, leave it like that. If, on the other hand, the placebo group is already doing better than the treatment group at the start, adjust for the baseline in your analysis.

IGNORE DROPOUTS

People who drop out of trials are statistically much more likely to have done badly and much more likely to have had side effects. They will only make your drug look bad. So ignore them: make no attempt to chase them up, do not include them in your final analysis.

CLEAN UP THE DATA

Look at your graphs. There will be some anomalous "outliers," or points that lie a long way from the others. If they are making your drug look bad, just delete them. But if they are helping your drug look good, even if they seem to be spurious results, leave them in.

"THE BEST OF FIVE ... NO ... SEVEN ... NO ... NINE!"

If the difference between your drug and placebo becomes significant four and a half months into a six-month trial, stop the trial immediately and start writing up the results; things might get less impressive if you carry on. Alternatively, if at six months the results are "nearly significant," extend the trial by another three months.

⌈ TORTURE THE DATA ⌉

If your results are bad, ask the computer to go back and see if any particular subgroups behaved differently. You might find that your drug works very well in Chinese women aged fifty-two to sixty-one. "Torture the data, and it will confess to anything," as they say at Guantánamo Bay.

TRY EVERY BUTTON ON THE COMPUTER

If you're really desperate and analyzing your data the way you planned does not give you the result you wanted, just run the figures through a wide selection of other statistical tests, even if they are entirely inappropriate, at random.

And when you're finished, the most important thing, of course, is to publish wisely. If you have a good trial, publish it in the biggest journal you can possibly manage. If you have a positive trial, but it was a completely unfair test, which will be obvious to everyone, then put it in an obscure journal (published, written, and edited entirely by the industry). Remember, the tricks we have just described hide nothing and will be obvious to anyone who reads your paper, but only if he or she reads it very attentively, so it's in your interest to make sure it isn't read beyond the abstract. Finally, if your finding is really embarrassing, hide it away somewhere, and cite "data on file." Nobody will know the methods, and it will be noticed only if someone comes pestering you for the data to do a systematic review. Hopefully, that won't be for ages.

HOW CAN THIS BE POSSIBLE?

When I explain this abuse of research to friends from outside medicine and academia, they are rightly amazed. "How can this be possible?" they say. Well, first, much bad research comes down to incompetence. Many of the methodological errors described above can come about by wishful thinking, as much as by mendacity. But is it possible to prove foul play?

On an individual level, it is sometimes quite hard to show that a trial has been deliberately rigged to give the right answer for its sponsors. Overall, however, the picture emerges very clearly. The issue has been studied so frequently that in 2003 a systematic review found thirty separate studies looking at whether funding in various groups of trials affected the findings. Overall, studies funded by a pharmaceutical company were found to be four times more likely to give results that were favorable to the company than were independent studies.

One review of bias tells a particularly Alice in Wonderland story. Fifty-six different trials comparing painkillers like ibupro-

fen, diclofenac, and so on were found. People often invent new
versions of these drugs in the hope that they might have fewer
side effects or be stronger (or stay in patent and make money). In
every single trial the sponsoring manufacturer's drug came out as
better than, or equal to, the others in the trial. On not one occasion did the manufacturer's drug come out worse. Philosophers and
mathematicians talk about "transitivity": if A is better than B,
and B is better than C, then C cannot be better than A. To put it
bluntly, this review of fifty-six trials exposed a singular absurdity:
all these drugs were better than one another.

But there is a surprise waiting around the corner. Astonishingly, when the methodological flaws in studies are examined, it
seems that industry-funded trials actually turn out to have *better*
research methods, on average, than independent trials. The most
that could be pinned on the drug companies were some fairly trivial howlers, things like using inadequate doses of the competitor's
drug (as we said above) or making claims in the conclusions section of the paper that exaggerated a positive finding. But these, at
least, were transparent flaws; you only had to read the trial to see
that the researchers had given a miserly dose of a painkiller, and
you should always read the methods and results section of a trial to
decide what its findings are, because the discussion and conclusion pages at the end are like the comment pages in a newspaper.
They're not where you get your news from.

How can we explain, then, the apparent fact that industry-
funded trials are so often so glowing? How can all the drugs simultaneously be better than all of the others? The crucial kludge may
happen after the trial is finished.

PUBLICATION BIAS AND SUPPRESSING NEGATIVE RESULTS

Publication bias is a very interesting and very human phenomenon. For a number of reasons, positive trials are more likely to get

published than negative ones. It's easy enough to understand, if you put yourself in the shoes of the researcher. First, when you get a negative result, it feels as if it's all been a bit of a waste of time. It's easy to convince yourself that you found nothing when in fact you discovered a very useful piece of information: that the thing you were testing *doesn't work*.

Rightly or wrongly, finding out that something doesn't work probably isn't going to win you a Nobel Prize—there's no justice in the world—so you might feel unmotivated about the project, or prioritize other projects ahead of writing up and submitting your negative finding to an academic journal, and so the data just sits, rotting, in your bottom drawer. Months pass. You get a new grant. The guilt niggles occasionally, but Monday's your day in the clinic, so Tuesday's the beginning of the week really, and there's the departmental meeting on Wednesday, so Thursday's the only day you can get any proper work done, because Friday's your teaching day, and before you know it, a year has passed, your supervisor retires, the new guy doesn't even know the experiment ever happened, and the negative trial data is forgotten forever, unpublished. If you are smiling in recognition at this paragraph, then you are a very bad person.

Even if you do get around to writing up your negative finding, it's hardly news. You're probably not going to get it into a big-name journal, unless it was a massive trial on something everybody thought was really whizbang until your negative trial came along and blew it out of the water, so as well as this being a good reason for you not bothering, it means the whole process will be heinously delayed: it can take a year for some of the slacker journals to reject a paper. Every time you submit to a different journal you might have to reformat the references (hours of tedium). If you aim too high and get a few rejections, it could be years until your paper comes out, even if you are being diligent; that's years of people not knowing about your study.

Publication bias is common, and in some fields it is more rife than in others. In 1995, only 1 percent of all articles published in alternative medicine journals gave a negative result. The most recent figure is 5 percent negative. This is very, very low, although to be fair, it could be worse. A review in 1998 looked at the entire canon of Chinese medical research and found that not one single negative trial had ever been published. Not one. You can see why I use CAM as a simple teaching tool for evidence-based medicine.

Generally the influence of publication bias is more subtle, and you can get a hint that publication bias exists in a field by doing something very clever called a funnel plot. This requires, only briefly, that you pay attention.

If there are lots of trials on a subject, then quite by chance they all will give slightly different answers, but you would expect them all to cluster fairly equally around the true answer. You would also expect that the bigger studies, with more participants in them, and with better methods, would be more closely clustered around the correct answer than the smaller studies; the smaller studies, meanwhile, will be all over the shop, unusually positive and negative at random, because in a study with, say, twenty patients, you need only three freak results to send the overall conclusions right off.

A funnel plot is a clever way of graphing this (as demonstrated by the graph on top of page 163). You put the effect (i.e., how effective the treatment is) on the x-axis, from left to right. Then, on the y-axis (top to bottom, for those of you who cut math) you put how big the trial was, or some other measure of how accurate it was. If there is no publication bias, you should see a nice inverted funnel. The big, accurate trials all cluster around one another at the top of the funnel, and then, as you go down the funnel, the little, inaccurate trials gradually spread out to the left and right, as they become more and more wildly inaccurate—both positively and negatively.

If there is publication bias, however, the results will be skewed. The smaller, more rubbish *negative* trials seem to be missing, be-

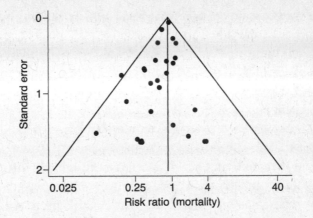

cause they were ignored—nobody had anything to lose by letting these tiny, unimpressive trials sit in their bottom drawer, and so only the positive ones were published. Not only has publication bias been demonstrated in many fields of medicine, but a paper has even found evidence of publication bias in studies of publication bias. The graph below is the funnel plot for that paper. This is what passes for humor in the world of evidence-based medicine. The most heinous recent case of publication bias has been in the area of SSRI antidepressant drugs, as has been shown in various papers. A group of academics published a paper in *The New England Journal of Medi-*

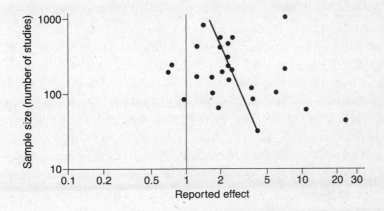

cine at the beginning of 2008 that listed all the trials on SSRIs that had ever been formally registered with the FDA, and examined the same trials in the academic literature. Thirty-seven studies were assessed by the FDA as positive: with one exception, every single one of those positive trials was properly written up and published. Meanwhile, twenty-two studies that had negative or iffy results were simply not published at all, and eleven were written up and published in a way that described them as having a positive outcome.

This is more than cheeky. Doctors need reliable information if they are to make helpful and safe decisions about prescribing drugs to their patients. Depriving them of this information, and deceiving them, are a major moral crime. If I weren't writing a light and humorous book about science right now, I would descend into gales of rage.

DUPLICATE PUBLICATION

Drug companies can go one better than neglecting negative studies. Sometimes, when they get positive results, instead of just publishing them once, they publish them several times, in different places, in different forms, so that it looks as if there were lots of different positive trials. This is particularly easy if you've performed a large "multicenter" trial, because you can publish overlapping bits and pieces from each center separately or in different permutations. It's also a very clever way of kludging the evidence, because it's almost impossible for the reader to spot.

A classic piece of detective work was performed in this area by a vigilant anesthetist from Oxford named Martin Tramer, who was looking at the efficacy of a nausea drug called ondansetron. He noticed that lots of the data in a meta-analysis he was doing seemed to be replicated; the results for many individual patients had been written up several times, in slightly different forms, in apparently different studies, in different journals. Crucially, data

that showed the drug in a better light was more likely to be dupli-
cated than the data that showed it to be less impressive, and overall, this led to a 23 percent overestimate of the drug's efficacy.

HIDING HARM

That's how drug companies dress up the positive results. What
about the darker, more headline-grabbing side, where they hide the
serious harms?

Side effects are a fact of life: they need to be accepted, managed in the context of benefits, and carefully monitored, because
the unintended consequences of interventions can be extremely serious. The stories that grab the headlines are ones in which there
is foul play, or a cover-up, but in fact, important findings can also be
missed for much more innocent reasons, like the normal human
processes of accidental neglect in publication bias or because the
worrying findings are buried from view in the noise of the data.

Antiarrhythmic drugs are an interesting example. People who
have heart attacks get irregular heart rhythms fairly commonly
(because bits of the timekeeping apparatus in the heart have been
damaged), and they also commonly die from them. Antiarrhythmic drugs are used to treat and prevent irregular rhythms in people who have them. Why not, thought doctors, just give them to
everyone who has had a heart attack? It made sense on paper, the
drugs seemed safe, and nobody knew at the time that they would
actually increase the risk of death in this group—because that
didn't make sense from the theory (like with antioxidants). But
they do, and at the peak of their use in the 1980s, antiarrhythmic
drugs were causing comparable numbers of deaths to the total
number of Americans who died in the Vietnam War. Information
that could have helped avert this disaster was sitting, tragically, in
a bottom drawer, as a researcher later explained: "When we carried out our study in 1980 we thought that the increased death

rate . . . was an effect of chance . . . The development of [the drug] was abandoned for commercial reasons, and so this study was therefore never published; it is now a good example of 'publication bias.' The results described here . . . might have provided an early warning of trouble ahead."

That was neglect, and wishful thinking. But sometimes it seems that dangerous effects from drugs can be either deliberately down-played or, worse than that, simply not published. There has been a string of major scandals from the pharmaceutical industry recently, in which it seems that evidence of harm for drugs including Vioxx and the SSRI antidepressants has gone missing in action. It didn't take long for the truth to out, and anybody who claims that these issues have been brushed under the medical carpet is simply igno-rant. They were dealt with, you'll remember, in the three highest-ranking papers in the *British Medical Journal*'s archive. They are worth looking at again, in more detail.

VIOXX

Vioxx was a painkiller developed by the Merck company and ap-proved by the American FDA in 1999. Many painkillers can cause gut problems—ulcers and more—and the hope was that this new drug might not have such side effects. This was examined in a trial called VIGOR, comparing Vioxx with an older drug, naproxen, and a lot of money was riding on the outcome. The trial had mixed results. Vioxx was no better at relieving the symptoms of rheuma-toid arthritis, but it did halve the risk of gastrointestinal events, which was excellent news. But an increased risk of heart attacks was also found.

When the VIGOR trial was published, however, this cardio-vascular risk was hard to see. There was an "interim analysis" for heart attacks and ulcers, in which ulcers were counted for longer than heart attacks. It wasn't described in the publication, and it overstated the advantage of Vioxx regarding ulcers, while under-stating the increased risk of heart attacks. "This untenable feature of trial design," said an unusually punishing editorial in *The New England Journal of Medicine*, "which inevitably skewed the results, was not disclosed to the editors or the academic authors of the study." Was it a problem? Yes. For one thing, three additional myo-cardial infarctions occurred in the Vioxx group in the month af-ter the researchers stopped counting, while none occurred in the naproxen control group.

An internal memo from Edward Scolnick, the company's chief scientist, shows that Merck knew about this cardiovascular risk ("It is a shame but it is a low incidence and it is mechanism based as we worried it was"). *The New England Journal of Medicine* was not im-pressed, publishing a pair of spectacularly critical editorials.

The worrying excess of heart attacks was only really picked up by people examining the FDA data, something that doctors tend, of course, not to do, as they read academic journal articles at best. In an attempt to explain the moderate extra risk of heart attacks that *could* be seen in the final paper, the authors proposed some-thing called the naproxen hypothesis: Vioxx wasn't causing heart attacks, they suggested, but naproxen was preventing them. There is no accepted evidence that naproxen has a strong protective ef-fect against heart attacks.

The internal memo, discussed at length in the coverage of the case, suggested that the company was concerned at the time. Even-tually more evidence of harm emerged. Vioxx was taken off the market in 2004, but analysts from the FDA estimated that it had caused between 88,000 and 139,000 heart attacks, 30 to 40 percent of which were probably fatal, in its five years on the market. It's hard to be sure if that figure is reliable, but when you look at the pattern of how the information came out, it's certainly felt, very widely, that both Merck and the FDA could have done much more to mitigate the damage done over the many years of this drug's life span, after the concerns were apparent to them. Data in medicine is important; it means lives. Merck has not admitted liability and has proposed a $4.85 billion settlement in the United States.

AUTHORS FORBIDDEN TO PUBLISH DATA

This all seems pretty bad. Which researchers are doing it, and why can't we stop them? Some, of course, are mendacious. But many have been bullied or pressured not to reveal information about the trials they have performed, funded by the pharmaceutical industry.

Here are two extreme examples of what is, tragically, a fairly common phenomenon. In 2000, a U.S. company filed a claim against both the lead investigators and their universities in an attempt to block publication of a study on an HIV vaccine that found the product was no better than placebo. The investigators felt they had to put patients before the product. The company felt otherwise. The results were published in *JAMA* that year.

In the second example, Nancy Olivieri, director of the Toronto Hemoglobinopathy Program, was conducting a clinical trial on deferiprone, a drug that removes excess iron from the bodies of patients who become iron overloaded after many blood transfusions. She was concerned when she saw that iron concentrations in the liver seemed to be poorly controlled in some of the patients, exceeding the safety threshold for increased risk of cardiac disease and early death. More extended studies suggested that deferiprone might accelerate the development of hepatic fibrosis.

The drug company, Apotex, threatened Olivieri, repeatedly and in writing, that if she published her findings and concerns, it would take legal action against her. With great courage—and, shamefully, without the support of her university—Olivieri presented her findings at several scientific meetings and in academic journals. She believed she had a duty to disclose her concerns, regardless of the personal consequences. It should never have been necessary for her to need to make that decision.

THE SINGLE CHEAP SOLUTION THAT WILL SOLVE ALL
THE PROBLEMS IN THE ENTIRE WORLD

What's truly extraordinary is that almost all these problems—the suppression of negative results, data dredging, hiding unhelpful data, and more—could largely be solved with one very simple intervention that would cost almost nothing: a clinical trials register, public, open, and properly enforced. This is how it would work. You're a drug company. Before you even start your study, you publish the protocol for it, the methods section of the paper, somewhere public. This means that everyone can see what you're going to do in your trial, what you're going to measure, how, in how many people, and so on, *before you start.*

The problems of publication bias, duplicate publication, and hidden data on side effects, which all cause unnecessary death and suffering, would be eradicated overnight, in one fell swoop. If you registered a trial, and conducted it, but it didn't appear in the literature, it would stick out like a sore thumb. Everyone, basically, would assume you had something to hide, because you probably would. There are trials registers at present, but they are a mess.

How much of a mess is illustrated by this last drug company ruse: "moving the goalposts." In 2002 Merck and Schering-Plough began a trial to look at ezetimibe, a drug to reduce cholesterol. They started out saying they were going to measure one thing as their test of whether the drug worked, but then announced, after the results were in, that they were going to count something else as the real test instead. This was spotted, and they were publicly rapped. Why? Because if you measure lots of things (as they did), some might be positive simply by chance. You cannot find your starting hypothesis in your final results. It makes the stats go all wonky.

ADVERTISEMENTS

Direct-to-consumer drug ads are properly bizarre, especially the TV ones. Your life is in disarray; your restless legs/migraine/cholesterol have taken over; all is panic; there is no sense anywhere. Then, when you take the right pill, suddenly the screen brightens up into a warm yellow, granny's laughing, the kids are laughing, the dog's tail is wagging, some nauseating child is playing with the hose on the lawn, spraying a rainbow of water into the sunshine while absolutely laughing his head off as all your relationships suddenly become successful again. All you have to do is "ask your doctor" and life will be good. It's worth noting that drug adverts aimed directly at the public are legally allowed only in the United States and New Zealand, as pretty much everywhere else in the developed world has banned them, for the simple reason that they work. Patients are so much more easily misled by drug company advertising than doctors that the budget for direct-to-consumer advertising in America has risen twice as fast as the budget for addressing doctors directly. These ads have been closely studied by medical academic researchers and have been repeatedly shown to increase patients' requests for the advertised drugs, as well as doctors' prescriptions for them. Even ads "raising awareness of a condition" under tighter Canadian regulations have been shown to double demand for a specific drug to treat that condition.

This is why drug companies are keen to sponsor patient groups, or to exploit the media for their campaigns, as has been seen recently in the news stories singing the praises of the breast cancer drug Herceptin or Alzheimer's drugs of borderline efficacy.

These advocacy groups demand vociferously in the media that the companies' drugs should be funded. I know people associated with these patient advocacy groups—academics— who have spoken out and tried to change their stance, without success: because in the case of the British Alzheimer's campaign in particular, it struck many people that the demands were rather one-sided. The National Institute for Clinical Excellence (NICE), which gives advice on whether drugs should be funded by the government, concluded that it couldn't justify paying for Alzheimer's drugs, partly because the evidence for their efficacy was weak and often looked only at soft, surrogate outcomes. The evidence for these drugs is often weak, because the drug companies have failed to subject their medications to sufficiently rigorous testing on real-world outcomes: rigorous testing that would be much less guaranteed to produce a positive result. Do patient organizations challenge the manufacturers to do better research? Do their members walk around with large placards campaigning against "surrogate outcomes in drugs research," demanding "More Fair Tests"? No. Oh, God. Everybody's bad. How did things get so awful?

10

WHY CLEVER PEOPLE
BELIEVE STUPID THINGS

The real purpose of the scientific method is to make sure nature hasn't misled you into thinking you know something you actually don't know.
—Robert Pirsig, *Zen and the Art of Motorcycle Maintenance*

Why do we have statistics, why do we measure things, and why do we count? If the scientific method has any authority—or, as I prefer to think of it, value—it is because it represents a systematic approach, but this is valuable only because the alternatives can be misleading. When we reason informally—call it intuition, if you like—we use rules of thumb that simplify problems for the sake of efficiency. Many of these shortcuts have been well characterized in a field called heuristics, and they are efficient ways of knowing in many circumstances.

This convenience comes at a cost—false beliefs—because there are systematic vulnerabilities in these truth-checking strategies that can be exploited. This is not dissimilar to the way that paintings can exploit shortcuts in our perceptual system: as objects become more distant, they appear smaller, and "perspective" can trick us into seeing three dimensions where there are only two, by taking advantage of this strategy used by our depth-checking apparatus. When our cognitive system—our truth-checking apparatus—is fooled, then, much like seeing depth in a flat painting, we come to erroneous conclusions about abstract things. We might misidentify normal fluctuations as meaningful patterns, for example, or ascribe causality where in fact there is none.

These are cognitive illusions, a parallel to optical illusions. They can be just as mind-boggling, and they cut to the core of why we do science, rather than base our beliefs on intuition informed by a "gist" of a subject acquired through popular media: because the world does not provide you with neatly tabulated data on interventions and outcomes. Instead it gives you random, piecemeal data in dribs and drabs over time, and trying to construct a broad understanding of the world from a memory of your own experiences would be like looking at the ceiling of the Sistine Chapel through a long, thin cardboard tube: you can try to remember the individual portions you've spotted here and there, but without a system and a model, you're never going to appreciate the whole picture.

Let's begin.

RANDOMNESS

As human beings we have an innate ability to make something out of nothing. We see shapes in the clouds and a man in the moon; gamblers are convinced that they have "runs of luck"; we take a perfectly cheerful heavy metal record, play it backward, and hear hidden messages about Satan. Our ability to spot patterns is what allows us to make sense of the world, but sometimes, in our eagerness, we are oversensitive and trigger-happy and mistakenly spot patterns where none exist.

In science, if you want to study a phenomenon, it is sometimes useful to reduce it to its simplest and most controlled form. There is a prevalent belief among sporting types that sportsmen, like gamblers (except more plausibly), have runs of luck. People ascribe this to confidence, "getting your eye in," "warming up," or more, and while it might exist in some games, statisticians have looked in various places where people have claimed it to exist and found no relationship between, say, hitting a home run in one inning, then hitting a home run in the next.

Because the "winning streak" is such a prevalent belief, it is an excellent model for looking at how we perceive random sequences of events. This was used by an American social psychologist named Thomas Gilovich in a classic experiment. He took basketball fans and showed them a random sequence of Xs and Os, explaining that they represented a player's hits and misses, and then asked them if they thought the sequences demonstrated streak shooting.

Here is a random sequence of figures from that experiment. You might think of it as being generated by a series of coin tosses.

OXXXOXXXOXXOOOXOOXXOO

The subjects in the experiment were convinced that this sequence exemplified streak shooting or runs of luck, and it's easy to see why, if you look again: six of the first eight shots were hits. No, wait: eight of the first eleven shots were hits. No way is that random . . .

What this ingenious experiment shows is how bad we are at correctly identifying random sequences. We are wrong about what they should look like: we expect too much alternation, so truly random sequences seem somehow too lumpy and ordered. Our intuitions about the most basic observation of all, distinguishing a pattern from mere random background noise, are deeply flawed.

This is our first lesson in the importance of using statistics instead of intuition. It's also an excellent demonstration of how strong the parallels are between these cognitive illusions and the perceptual illusions with which we are more familiar. You can stare at a visual illusion all you like, talk or think about it, but it will still look "wrong." Similarly, you can look at that random sequence above as hard as you like: it will still look lumpy and ordered, in defiance of what you now know.

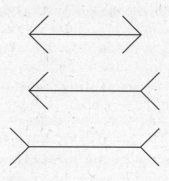

REGRESSION TO THE MEAN

We have already looked at regression to the mean in our section on homeopathy; it is the phenomenon whereby when things are at their extremes, they are likely to settle back down to the middle, or regress to the mean.

We saw this with reference to the *Sports Illustrated* jinx, but also applied it to the matter in hand, the question of people getting better; we discussed how people will do something when their back pain is at its worst—visit a homeopath, perhaps—and how although it was going to get better anyway (because when things are at their worst, they generally do), they ascribe their improvement to the treatment.

There are two discrete things happening when we fall prey to this failure of intuition. First, we have failed to spot correctly the pattern of regression to the mean. Second crucially, we have then decided that something must have *caused* this illusory pattern: specifically, a homeopathic remedy, for example. Simple regression is confused with causation, and this is perhaps quite natural for animals like humans, whose success in the world depends on our being able to spot causal relationships rapidly and intuitively: we are inherently oversensitive to them.

To an extent, when we discussed the subject earlier, I relied on your goodwill and on the likelihood that from your own experience you could agree that this explanation made sense. But it has been demonstrated in another ingeniously pared-down experiment, in which all the variables were controlled, but people still saw a pattern and causality where there was none.

The subjects in the experiment played the role of a teacher trying to make a child arrive punctually at school for 8:30 a.m. They sat at a computer on which it appeared that each day, for fifteen consecutive days, the supposed child would arrive sometime between 8:20 and 8:40, but unbeknownst to the subjects, the arrival times were entirely random and predetermined before the experiment began. Nonetheless, all the subjects were allowed to use punishments for lateness and rewards for punctuality, in whatever permutation they wished. When they were asked at the end to rate their strategy, 70 percent concluded that reprimand was more effective than reward in producing punctuality from the child.

These subjects were convinced that their intervention had an effect on the punctuality of the child, despite the child's arrival time being entirely random and exemplifying nothing more than regression to the mean. By the same token, when homeopathy has been shown to elicit no more improvement than placebo, people are still convinced that it has a beneficial effect on their health.

To recap:

1. We see patterns where there is only random noise.
2. We see causal relationships where there are none.

These are two very good reasons to measure things formally. It's bad news for intuition already. Can it get much worse?

THE BIAS TOWARD POSITIVE EVIDENCE

> It is the peculiar and perpetual error of the human understanding to be more moved and excited by affirmatives than negatives.
>
> —Francis Bacon

It gets worse. It seems we have an innate tendency to seek out and overvalue evidence that confirms a given hypothesis. To try to remove this phenomenon from the controversial arena of CAM—or the MMR scare, which is where this is headed—we are lucky to have more pared-down experiments that illustrate the general point.

Imagine a table with four cards on it, marked "A," "B," "2," and "3." Each card has a letter on one side and a number on the other. Your task is to determine whether all cards with a vowel on one side have an even number on the other. Which two cards would you turn over? Everybody chooses the "A" card, obviously, but like many people—unless you really forced yourself to think hard about it—you would probably choose to turn over the "2" card as well. That's because these are the cards that would produce information *consistent* with the hypothesis you are supposed to be testing. But in fact, the cards you need to flip are the "A" and the "3," because finding a vowel on the back of the "2" would tell you nothing about "all cards," it would just confirm "some cards," whereas finding a vowel on the back of "3" would comprehensively disprove your hypothesis. This modest brainteaser demonstrates our tendency, in our unchecked intuitive reasoning style, to seek out information that confirms a hypothesis, and it demonstrates the phenomenon in a value-neutral situation.

This same bias in seeking out confirmatory information has been demonstrated in more sophisticated social psychology experiments. When trying to determine if someone is an "extrovert," for example, many subjects will ask questions for which a positive answer would confirm the hypothesis ("Do you like going to parties?") rather than refute it.

We show a similar bias when we interrogate information from our own memory. In one experiment, subjects first read a vignette about a woman who exemplified various introverted and extroverted behaviors and then were divided into two groups. One group was asked to consider the woman's suitability for a job as a librarian, while the other was asked to consider her suitability for a job as a real estate agent. Both groups were asked to come up with examples of both her extroversion and her introversion. The group considering her for the librarian job recalled more examples of introverted behavior, while the group considering her for a job selling real estate cited more examples of extroverted behavior.

This tendency is dangerous, because if you ask only questions that confirm your hypothesis, you will be more likely to elicit information that confirms it, giving a spurious sense of confirmation. It also means—if we think more broadly—that the people who pose the questions already have a head start in popular discourse.

So we can add to our running list of cognitive illusions, biases, and failings of intuition:

3. We overvalue confirmatory information for any given hypothesis.
4. We seek out confirmatory information for any given hypothesis.

BIASED BY OUR PRIOR BELIEFS

[I] followed a golden rule, whenever a new observation or thought came across me, which was opposed to my general results, to make a

memorandum of it without fail and at once; for I had found by experience that such facts and thoughts were far more apt to escape from
the memory than favorable ones.   —Charles Darwin

This is the reasoning flaw that everybody does know about, and even
if it's the least interesting cognitive illusion—because it's an obvious
one—it has been demonstrated in experiments that are so close to
the bone that you may find them, as I do, quite unnerving.

The classic demonstration of people's being biased by their
prior beliefs comes from a study looking at beliefs about the death
penalty. A large number of proponents and opponents of state ex-
ecutions were collected. They all were shown two pieces of evidence on the deterrent effect of capital punishment: one supporting
a deterrent effect, the other providing evidence against it.

The evidence they were shown was as follows:

- A comparison of murder rates in one U.S. state before
  the death penalty was brought in, and after.
- A comparison of murder rates in different states, some
  with and some without the death penalty.

But there was a very clever twist. The proponents and opponents
of capital punishment were each further divided into two smaller
groups. So overall, half the proponents and opponents of capital
punishment had their opinions reinforced by before/after data but
challenged by state/state data, and vice versa.

Asked about the evidence, the subjects confidently uncovered
flaws in the methods of the research that went against their preexisting view, but downplayed the flaws in the research that supported their view. Half the proponents of capital punishment, for
example, picked holes in the idea of state/state comparison data,
on methodological grounds, because that was the data that went
against their view, while they were happy with the before/after

data; but the other half of the proponents of capital punishment rubbished the before/after data, because in their case they had been exposed to before/after data that challenged their view and state/state data that supported it.

Put simply, the subjects' faith in research data was not predicated on an objective appraisal of the research methodology, but on whether the results validated their preexisting views. This phenomenon reaches its pinnacle in alternative therapists—or scaremongers—who unquestioningly champion anecdotal data, while meticulously examining every large, carefully conducted study on the same subject for any small chink that would permit them to dismiss it entirely.

This, once again, is why it is so important that we have clear strategies available to us to appraise evidence, regardless of its conclusions, and this is the major strength of science. In a systematic review of the scientific literature, investigators will sometimes mark the quality of the "methods" section of a study blindly—that is, without looking at the "results" section—so that it cannot bias their appraisal. Similarly, in medical research there is a hierarchy of evidence: a well-performed trial is more significant than survey data in most contexts, and so on.

So we can add to our list of new insights about the flaws in intuition:

5. Our assessment of the quality of new evidence is biased by our previous beliefs.

AVAILABILITY

We spend our lives spotting patterns and picking out the exceptional and interesting things. You don't waste cognitive effort, every time you walk into your house, noticing and analyzing all the many features in the visually dense environment of your kitchen. You do notice the broken window and the missing television.

When information is made more "available," as psychologists call it, it becomes disproportionately prominent. There are a number of ways this can happen, and you can pick up a picture of them from a few famous psychology experiments into the phenomenon.

In one, subjects were read a list of male and female names, in equal number, and then asked at the end whether there were more men or women in the list. When the men in the list had names like Ronald Reagan, but the women were unheard of, people tended to answer that there were more men than women, and vice versa.

Our attention is always drawn to the exceptional and the interesting, and if you have something to sell, it makes sense to guide people's attention to the features you most want them to notice. When slot machines pay up, they make a theatrical "kerchunk-kerchunk" sound with every coin they spit out, so that everybody in the casino can hear it, but when you lose, they don't draw attention to themselves. Lottery companies, similarly, do their absolute best to get their winners prominently into the media, but it goes without saying that you, as a lottery loser, have never had your outcome paraded for the TV cameras.

As we shall see, the tragic anecdotes about the MMR vaccine are disproportionately misleading, not just because the statistical context is missing, but because of their "high availability": they are dramatic, associated with strong emotion, and amenable to strong visual imagery. They are concrete and memorable, rather than abstract. No matter what you do with statistics about risk or recovery, your numbers will always have inherently low psychological availability, unlike miracle cures, scare stories, and distressed parents.

It's because of availability, and our vulnerability to drama, that people are more afraid of sharks at the beach, or of fairground rides on the pier, than they are of flying to Florida or driving to the coast. This phenomenon is even demonstrated in patterns of smoking cessation among doctors. You'd imagine, since they are rational actors, that all doctors would simultaneously have seen sense and stopped smoking once they'd read the studies showing the phenomenally compelling relationship between cigarettes and lung cancer. These are men of applied science, after all, who are able, every day, to translate cold statistics into meaningful information and beating human hearts.

But in fact, from the start, doctors working in specialties like chest medicine and oncology, where they witnessed patients dying of lung cancer with their own eyes, were proportionately more likely to give up cigarettes than their colleagues in other specialties. Being shielded from the emotional immediacy and drama of consequences matters.

SOCIAL INFLUENCES

Last in our whistle-stop tour of irrationality comes our most self-evident flaw. It feels almost too obvious to mention, but our values are socially reinforced by conformity and by the company we keep. We are selectively exposed to information that revalidates our beliefs, partly because we expose ourselves to *situations* in which those beliefs are apparently confirmed; partly because we ask questions that will—by their very nature, for the reasons described above—give validating answers; and partly because we selectively expose ourselves to *people* who validate our beliefs.

It's easy to forget the phenomenal impact of conformity. You doubtless think of yourself as a fairly independent-minded person, and you know what you think. I would suggest that the same beliefs were held by the subjects of Solomon Asch's experiments into social conformity. These subjects were placed near one end of a line of actors who presented themselves as fellow experimental subjects but were actually in cahoots with the experimenters. Cards were held up with one line marked on each of them, and then another card was held up with three lines of different lengths: six inches, eight inches, ten inches.

Everyone called out in turn which line on the second card was the same length as the line on the first. For six of the eighteen pairs of cards the accomplices gave the correct answer, but for the other twelve they called out the wrong answer. In all but a quarter of the cases, the experimental subjects went along with the incorrect answer from the crowd of accomplices on one or more occasions, defying the clear evidence of their own senses.

That's an extreme example of conformity, but the phenomenon is all around us. Communal reinforcement is the process by which a claim becomes a strong belief, through repeated assertion by members of a community. The process is independent of whether the claim has been properly researched or is supported by empirical data significant enough to warrant belief by reasonable people.

Communal reinforcement goes a long way toward explaining how religious beliefs can be passed on in communities from generation to generation. It also explains how testimonials within communities of therapists, psychologists, celebrities, theologians, politicians, talk show hosts, and so on can supplant and become more powerful than scientific evidence.

> When people learn no tools of judgment and merely follow their hopes, the seeds of political manipulation are sown.
>
> —Stephen Jay Gould

There are many other well-researched areas of bias. We have a disproportionately high opinion of ourselves, which is nice. A large majority of the public think they are more fair-minded, less prejudiced, more intelligent, and more skilled at driving than the average person, when of course, only half of us can be better than the median.* Most of us exhibit something called attributional bias: we believe our successes are due to our own internal faculties, and our failures are due to external factors; whereas for others, we believe their successes are due to luck, and their failures to their own flaws. We can't all be right.

Last, we use context and expectation to bias our appreciation of a situation, because in fact, that's the only way we can think. Artificial intelligence research has drawn a blank so far largely because of something called the frame problem: you can tell a computer how to process information, and give it all the information in the world, but as soon as you give it a real-world problem—a sentence to understand and respond to, for example—computers perform much worse than we might expect, because they don't know what information is relevant to the problem. This is something humans are very good at—filtering irrelevant information—but that skill comes at a cost of ascribing disproportionate bias to some contextual data.

We tend to assume, for example, that positive characteristics cluster: people who are attractive must also be good; people who seem kind might also be intelligent and well informed. Even this has been demonstrated experimentally: identical essays in neat handwriting score higher than messy ones, and the behavior of sporting teams that wear black is rated as more aggressive and unfair than teams that wear white.

And no matter how hard you try, sometimes things just are very counterintuitive, especially in science. Imagine there are twenty-three people in a room. What is the chance that two of them celebrate their birthday on the same date? One in two.†

---

*I'd be genuinely intrigued to know how long it takes to find someone who can tell you the difference between "median," "mean," and "mode," from where you are sitting right now.
†If it helps to make this feel a bit more plausible, bear in mind that you only need any two dates to coincide. With forty-seven people, the probability increases to 0.95; that's

When it comes to thinking about the world around you, you have a range of tools available. Intuitions are valuable for all kinds of things, especially in the social domain: deciding if your girlfriend is cheating on you, perhaps, or whether a business partner is trustworthy. But for mathematical issues, or assessing causal relationships, intuitions are often completely wrong, because they rely on shortcuts that have arisen as handy ways to solve complex cognitive problems rapidly, but at a cost of inaccuracies, misfires, and oversensitivity.

It's not safe to let our intuitions and prejudices run unchecked and unexamined: it's in our interest to challenge these flaws in intuitive reasoning wherever we can, and the methods of science and statistics grew up specifically in opposition to these flaws. Their thoughtful application is our best weapon against these pitfalls, and the challenge, perhaps, is to work out which tools to use where. Because trying to be "scientific" about your relationship with your partner is as stupid as following your intuitions about causality.

Now let's see how journalists deal with stats.

---

nineteen times out of twenty! (Fifty-seven people and it's 0.99; seventy people and it's 0.999.) This is beyond your intuition; at first glance, it makes no sense at all.

11

BAD STATS

Now that you appreciate the value of statistics—the benefits and risks of intuition—we can look at how these numbers and calculations are repeatedly misused and misunderstood. Our first examples will come from the world of journalism, but the true horror is that journalists are not the only ones to make basic errors of reasoning.

Numbers, as we shall see, can ruin lives.

THE BIGGEST STATISTIC

Newspapers like big numbers and eye-catching headlines. They need miracle cures and hidden scares, and small percentage shifts in risk will never be enough for them to sell readers to advertisers (because that is the business model). To this end they pick the single most melodramatic and misleading way of describing any statistical increase in risk, which is called the relative risk increase.

Let's say the risk of having a heart attack in your fifties is 50 percent higher if you have high cholesterol. That sounds pretty bad. Let's say the extra risk of having a heart attack if you have high cholesterol is only 2 percent. That sounds OK to me. But they're the same (hypothetical) figures. Let's try this. Out of a hundred men in their fifties with normal cholesterol, four will be expected to have a heart attack, whereas out of a hundred men with high cholesterol, six will be expected to have a heart attack. That's two extra heart attacks per hundred. Those are called natural frequencies.

Natural frequencies are readily understandable, because instead of using probabilities, or percentages, or anything even slightly technical or difficult, they use concrete numbers, just like the ones you use every day to check if you've lost a kid on a bus trip or got the right change in a shop. Lots of people have argued that we evolved to reason and do math with concrete numbers like these, and not with probabilities, so we find them more intuitive. Simple numbers are simple.

The other methods of describing the increase have names too. From our example above, with high cholesterol, you could have a 50 percent increase in risk (the "relative risk increase"), or a 2 percent increase in risk (the "absolute risk increase"), or, let me ram it home, the easy one, the informative one, an extra two heart attacks for every hundred men, the natural frequency.

As well as being the most comprehensible option, natural frequencies contain more information than the journalists' relative risk increase. Recently, for example, we were told that red meat causes bowel cancer, and ibuprofen increases the risk of heart attacks; but if you followed the news reports, you would be no wiser. Try this, on bowel cancer, from the *Today* program on Radio 4: "A bigger risk meaning what, Professor Bingham?" "A third higher risk." "That sounds an awful lot, a third higher risk; what are we talking about in terms of numbers here?" "A difference . . . of around about twenty people per year." "So it's still a small number?" "Umm . . . per 10,000 . . ."

These things are hard to communicate if you step outside the simplest format. Professor Sheila Bingham was the director of the MRC Centre for Nutritional Epidemiology in Cancer Prevention and Survival at the University of Cambridge and dealt with these numbers for a living, but in this (entirely forgivable) fumbling on a live radio show she was not alone; there are studies of doctors, and commissioning committees for local health authorities, and members of the legal profession that show that people who inter-pret and manage risk for a living often have huge difficulties ex-pressing on the spot what they mean. They are also much more likely to make the right decisions when information about risk is presented as natural frequencies, rather than as probabilities or percentages.

For painkillers and heart attacks, another front-page story, the desperate urge to choose the biggest possible number led to the figures being completely inaccurate in many newspapers. The re-ports were based on a study that had observed participants over four years, and the results suggested, using natural frequencies, that you would expect one extra heart attack for every 1,005 peo-ple taking ibuprofen. Or as the *Daily·Mail*, in an article titled "How Pills for Your Headache Could Kill," reported: "British re-search revealed that patients taking ibuprofen to treat arthritis face a 24 percent increased risk of suffering a heart attack." Feel the fear.

Almost everyone reported the relative risk increases: diclofenac increases the risk of heart attack by 55 percent; ibuprofen, by 24 per-cent. *The Boston Globe* was clever enough to report the natural frequencies: 1 extra heart attack in 1,005 people on ibuprofen. The U.K.'s *Daily Mirror*, meanwhile, tried and failed, reporting that 1 in 1,005 people on ibuprofen "will suffer heart failure over the follow-ing year." No. It's heart attack, not heart failure, and it's 1 *extra* per-son in 1,005, on top of the heart attacks you'd get anyway. Several other papers repeated the same mistake.

Often it's the fault of the press releases, and academics can themselves be as guilty as the rest when it comes to overdramatizing their research. But if anyone in a position of power is reading this, here is the information I would like from a newspaper, to help me make decisions about my health, when reporting on a risk: I want to know whom you're talking about (e.g., men in their fifties); I want to know what the baseline risk is (e.g., four men out of a hundred will have a heart attack over ten years); and I want to know what the increase in risk is, as a natural frequency (two extra men out of that hundred will have a heart attack over ten years). I also want to know exactly what's causing that increase in risk: an occasional headache pill, or a daily tubful of pain-relieving medication for arthritis. Then I will consider reading your newspapers again, instead of blogs that are written by people who understand research and that link reliably back to the original academic paper, so that I can double-check their précis when I wish.

More than a hundred years ago, H. G. Wells said that statistical thinking would one day be as important as the ability to read and write in a modern technological society. I disagree; probabilistic reasoning is difficult for everyone, but everyone understands normal numbers. This is why natural frequencies are the only sensible way to communicate risk.

CHOOSING YOUR FIGURES

Sometimes the mispresentation of figures goes so far beyond reality that you can only assume mendacity. Often these situations seem to involve morality: drugs, abortion, and the rest. With very careful selection of numbers, in what some might consider to be a cynical and immoral manipulation of the facts for personal gain, you can sometimes make figures say anything you want.

The UK's *Independent* was in favor of legalizing cannabis for many years, but in March 2007 it decided to change its stance. One option would have been simply to explain this as a change of heart, or a reconsideration of the moral issues. Instead it was decorated with science—as cowardly zealots have done from eugenics through to prohibition—and justified with a fictitious change in the facts. CANNABIS—AN APOLOGY was the headline for its front-page splash: "In 1997, this newspaper launched a campaign to decriminalise the drug. If only we had known then what we can reveal today . . . Record numbers of teenagers are requiring drug treatment as a result of smoking skunk, the highly potent cannabis strain that is 25 times stronger than resin sold a decade ago." Twice in this story we are told that cannabis is twenty-five times stronger than it was a decade ago. For the paper's former editor Rosie Boycott, in her melodramatic recantation, skunk was "thirty times stronger." In one inside feature the strength issue was briefly downgraded to a "can be." The paper even referenced its figures: "The Forensic Science Service says that in the early Nineties cannabis would contain around 1 per cent tetrahydrocannabinol (THC), the mind-altering compound, but can now have up to 25 percent."

This is all sheer fantasy.

I've got the U.K.'s Forensic Science Service data right here in front of me, and the earlier data from the Laboratory of the Government Chemist, the United Nations Drug Control Program, and the European Monitoring Centre for Drugs and Drug Addiction. I'm going to share it with you, because I happen to think that people are very well able to make their own minds up about important social and moral issues when given the facts.

The data from the Laboratory of the Government Chemist goes from 1975 to 1989. Cannabis resin pootles around between 6 percent and 10 percent THC, herbal between 4 percent and 6 percent. There is no clear trend.

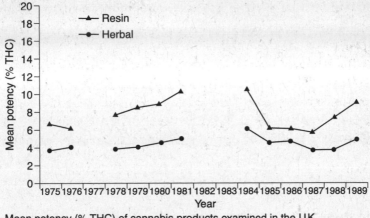

Mean potency (% THC) of cannabis products examined in the U.K.
(Laboratory of the Government Chemist, 1975–89)

The Forensic Science Service data then takes over to produce the more modern figures, showing not much change in resin and domestically produced indoor herbal cannabis doubling in potency from 6 percent to around 12 or 14 percent (2003–05 data in table under references).

The rising trend of cannabis potency is gradual, fairly unspectacular, and driven largely by the increased availability of domestic, intensively grown indoor herbal cannabis.

"Twenty-five times stronger," remember. Repeatedly, and on the front page.

If you were in the mood to quibble with *The Independent*'s moral and political reasoning, as well as its evident and shameless venality, you could argue that intensive indoor cultivation of a plant that grows perfectly well outdoors is the cannabis industry's reaction to the product's illegality itself. It is dangerous to import cannabis in large amounts. It is dangerous to be caught growing a field of it. So it makes more sense to grow it intensively indoors, using expensive real estate, but producing a more concentrated drug. More concentrated drugs products are, after all, a natural conse-

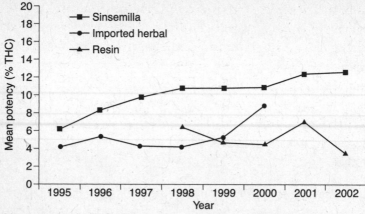

Mean potency (% THC) of cannabis products examined in the U.K.
(Forensic Science Service, 1995–2002)

Year	Sinsemilla %	Resin %	"Traditional" imported herbal %
1995	5.8	No data	3.9
1996	8.0	No data	5.0
1997	9.4	No data	4.0
1998	10.5	6.1	3.9
1999	10.6	4.4	5.0
2000	12.2	4.2	8.5
2001	12.3	6.7	No data
2002	12.3	3.2	No data
2003	12.0	4.6	No data
2004	12.7	1.6	No data
2005	14.2	6.6	No data

Mean THC content of cannabis products seized in the U.K.
(Forensic Science Service, 1995–2005)

quence of illegality. You can't buy coca leaves in South London, but you can buy crack.

There is, of course, exceptionally strong cannabis to be found in some parts of the British market today, but then there always has been. To get its scare figure, *The Independent* can only have compared the *worst* cannabis from the past with the *best* cannabis of today. It's an absurd thing to do, and moreover, you could have cooked the books in exactly the same way thirty years ago if you'd wanted; the figures for individual samples are available, and in 1975 the weakest herbal cannabis analyzed was 0.2 percent THC, while in 1978 the strongest herbal cannabis was 12 percent. By these figures, in just three years herbal cannabis became "sixty times stronger."

And this scare isn't even new. In the mid-1980s, during Ronald Reagan's "war on drugs," American campaigners were claiming that cannabis was fourteen times stronger than in 1970. Which sets you thinking. If it was fourteen times stronger in 1986 than in 1970, and it's twenty-five times stronger today than at the beginning of the 1990s, does that mean it's now 350 times stronger than in 1970?

That's not even a crystal in a plant pot. It's impossible. It would require more THC to be present in the plant than the total volume of space taken up by the plant itself. It would require matter to be condensed into superdense quark-gluon plasma cannabis. For God's sake don't tell the newspapers such a thing is possible.

COCAINE FLOODS THE PLAYGROUND

We are now ready to move on to some more interesting statistical issues, with another story from an emotive area, an article in *The Times* (London) in March 2006 headed: COCAINE FLOODS THE PLAYGROUND. "Use of the addictive drug by children doubles in a year," said the subheading. Was this true?

If you read the press release for the government survey on which the story is based, it reports "almost no change in patterns of drug use, drinking or smoking since 2000." But this was a government press release, and journalists are paid to investigate: perhaps the press release was hiding something, to cover up for government failures. The *Telegraph* also ran the "cocaine use doubles" story, and so did the *Mirror*. Did the journalists find the news themselves, buried in the report?

You can download the full document online. It's a survey of nine thousand children, aged eleven to fifteen, in 305 schools. The three-page summary said, again, that there was no change in prevalence of drug use. If you look at the full report, you will find the raw data tables: when asked whether they had used cocaine in the past year, 1 percent said yes in 2004, and 2 percent said yes in 2005.

So the newspapers were right: it doubled? No. Almost all the figures given were 1 percent or 2 percent. They'd all been rounded off. Civil servants are very helpful when you ring them up. The actual figures were 1.4 percent for 2004 and 1.9 percent for 2005, not 1 percent and 2 percent. So cocaine use hadn't doubled at all. But people were still eager to defend this story; cocaine use, after all, had increased, yes?

No. What we now have is a relative risk increase of 35.7 percent, or an absolute risk increase of 0.5 percent. If we use the real numbers, out of nine thousand kids we have about forty-five more saying yes to the question "Did you take cocaine in the past year?"

Presented with a small increase like this, you have to think: Is it statistically significant? I did the math, and the answer is yes, it is, in that you get a p-value of less than 0.05. What does "statistically significant" mean? It's just a way of expressing the likelihood that the result you got was attributable merely to chance. Sometimes you might throw heads five times in a row, with a completely normal coin, especially if you kept tossing it for long enough. Imagine a jar of 980 blue marbles, and 20 red ones, all mixed up; every now and then—albeit rarely—picking blindfolded, you might pull out 3 red ones in a row, just by chance. The standard cutoff point for statistical significance is a p-value of 0.05, which is just another way of saying, "If I did this experiment a hundred times, I'd expect a spurious positive result on five occasions, just by chance."

To go back to our concrete example of the kids in the playground, let's imagine that there was definitely no difference in cocaine use, but you conducted the same survey a hundred times. You might get a difference like the one we have seen here, just by chance, just because you randomly happened to pick up more of the kids who had taken cocaine this time around. But you would expect this to happen less than five times out of your hundred surveys.

So we have a risk increase of 35.7 percent, which seems at face value to be statistically significant; but it is an isolated figure. To "data mine," taking it out of its real-world context and saying it is significant, is misleading. The statistical test for significance assumes that every data point is independent, but here the data is "clustered," as statisticians say. They are not data points; they are real children, in 305 schools. They hang out together; they copy one another; they buy drugs from one another; there are crazes, epidemics, group interactions.

The increase of forty-five kids taking cocaine could have been a massive epidemic of cocaine use in one school, or a few groups of a dozen kids in a few different schools, or miniepidemics in a handful of schools. Or forty-five kids independently sourcing and consuming cocaine alone without their friends, which seems pretty unlikely to me.

This immediately makes our increase less statistically significant. The small increase of 0.5 percent was significant only because it came from a large sample of nine thousand data points—like nine thousand tosses of a coin—and the one thing almost everyone knows about studies like this is that a bigger sample size means the results are probably more significant. But if they're not independent data points, then you have to treat it, in some respects, like a smaller sample, so the results become less significant. As statisticians would say, you must "correct for clustering." This is done with clever math that makes everyone's head hurt. All you need to know is that the reasons why you must correct for clustering are transparent, obvious, and easy, as we have just seen (in fact, as with many implements, knowing when to use a statistical tool is a different and equally important skill from understanding how it is built). When you correct for clustering, you greatly reduce the significance of the results. Will our increase in cocaine use, already down from "doubled" to "35.7 percent," even survive?

No. Because there is a final problem with this data: there is so much of it to choose from. There are dozens of data points in the report: on solvents, cigarettes, ketamine, cannabis, and so on. It is standard practice in research that we only accept a finding as significant if it has a p-value of 0.05 or less. But as we said, a p-value of 0.05 means that for every hundred comparisons you do, five will be positive by chance alone. From this report you could have done dozens of comparisons, and some of them would indeed have shown increases in usage—but by chance alone, and the cocaine figure could be one of those. If you roll a pair of dice often enough, you will get a double six three times in a row on many occasions. This is why statisticians do a "correction for multiple comparisons," a correction for "rolling the dice" lots of times. This, like correcting for clustering, is particularly brutal on the data and often reduces the significance of findings dramatically.

Data dredging is a dangerous profession. You could—at face value, knowing nothing about how stats works—have said that this government report showed a significant increase of 35.7 percent in cocaine use. But the stats nerds who compiled it knew about clustering and Bonferroni's correction for multiple comparisons. They are not stupid; they do stats for a living.

That, presumably, is why they said quite clearly in their summary, in their press release, and in the full report that there was no change from 2004 to 2005. But the journalists did not want to believe this. They tried to reinterpret the data for themselves; they looked under the hood, and they thought they'd found the news. The increase went from 0.5 percent—a figure that might be a gradual trend, but could equally well be an entirely chance finding—to a front-page story in *The Times* (London) about cocaine use's doubling. You might not trust the press release, but if you don't know about numbers, then you take a big chance when you delve under the hood of a study to find a story.

OK, BACK TO AN EASY ONE

There are also some perfectly simple ways to generate ridiculous statistics, and two common favorites are to select an unusual sample group of people and to ask them a stupid question. Let's say 70 percent of all women want Prince Charles to be told to stop interfering in public life. Oh, hang on—70 percent of all women *who visit my website* want Prince Charles to be told to stop interfering in public life. You can see where we're going. And of course, in surveys, if they are voluntary, there is something called selection bias; only the people who can be bothered to fill out the survey form will actually have a vote registered.

There was an excellent example of this in *The Daily Telegraph* in the last days of 2007. DOCTORS SAY NO TO ABORTIONS IN THEIR SURGERIES was the headline. "Family doctors are threatening a revolt against government plans to allow them to perform abortions in their surgeries, the *Daily Telegraph* can disclose." A revolt? "Four out of five doctors do not want to carry out terminations even though the idea is being tested in NHS [National Health Service] pilot schemes, a survey has revealed."

Where did these figures come from? A systematic survey of all doctors, with lots of chasing to catch the nonresponders? Telephoning them at work? A postal survey, at least? No. It was an online vote on a doctors' chat site that produced this major news story. Here is the question and the options given:

"[Doctors] should carry out abortions in their surgeries"
    Strongly agree, agree, don't know, disagree, strongly disagree.

We should be clear: I myself do not fully understand this question. Is that "should" as in "should"? As in "ought to"? And in what circumstances? With extra training, time, and money? With extra systems in place for adverse outcomes? And remember, this is a website where doctors—bless them—go to moan. Are they just saying no because they're grumbling about more work and low morale?

More than that, what exactly does "abortion" mean here? Looking at the comments in the chat forum, I can tell you that plenty of the doctors seemed to think it was about surgical abortions, not the relatively safe oral pill for termination of pregnancy. Doctors aren't that bright, you see. Here are some quotes:

This is a preposterous idea. How can [doctors] ever carry out abortions in their own surgeries. What if there was a major complication like uterine and bowel perforation?

[Doctor's] surgeries are the places par excellence where infective disorders present. The idea of undertaking there any sort of sterile procedure involving an abdominal organ is anathema.

The only way it would or rather should happen is if [doctor] practices have a surgical day care facility as part of their premises which is staffed by appropriately trained staff, i.e. theater staff, anesthetist and gynecologist . . . any surgical operation is not without its risks, and presumably

[we] will undergo gynecological surgical training in order
to perform.

What are we all going on about? Let's all carry out abortions in our surgeries, living rooms, kitchens, garages, corner shops, you know, just like in the old days.

And here's my favorite:

I think that the question is poorly worded and I hope that
[the doctors' website] do[es] not release the results of this
poll to *The Daily Telegraph*.

BEATING YOU UP

It would be wrong to assume that the kinds of oversights we've
covered so far are limited to the lower echelons of society, like doctors and journalists. Some of the most sobering examples come from
the very top.

In 2006, after a major British government report, the media reported that one murder a week is committed by someone with psychiatric problems. Psychiatrists should do better, the newspapers
told us, and prevent more of these murders. All of us would agree,
I'm sure, with any sensible measure to improve risk management
and violence, and it's always timely to have a public debate about
the ethics of detaining psychiatric patients (although in the name
of fairness I'd like to see preventive detention discussed for all other
potentially risky groups too—like alcoholics, the repeatedly violent,
people who have abused staff in the job center, and so on).

But to engage in this discussion, you need to understand the
math of predicting very rare events. Let's take a very concrete example and look at the HIV test. What features of any diagnostic
procedure do we measure in order to judge how useful it might be?

Statisticians would say the blood test for HIV has a very high "sensitivity," at 0.999. That means that if you do have the virus, there is a 99.9 percent chance that the blood test will be positive. They would also say the test has a high "specificity" of 0.9999, so if you are not infected, there is a 99.99 percent chance that the test will be negative. What a smashing blood test.*

But if you look at it from the perspective of the person being tested, the math gets slightly counterintuitive. Because weirdly, the meaning, the predictive value, of an individual's positive or negative test is changed in different situations, depending on the background rarity of the event that the test is trying to detect. The rarer the event in your population, the worse your test becomes, even though it is the same test.

This is easier to understand with concrete figures. Let's say the HIV infection rate among high-risk men in a particular area is 1.5 percent. We use our excellent blood test on 10,000 of these men, and we can expect 151 positive blood results overall: 150 will be our truly HIV positive men, who will get true positive blood tests, and 1 will be the one false positive we could expect from having 10,000 HIV negative men being given a test that is wrong one time in 10,000. So, if you get a positive HIV blood test result, in these circumstances your chances of being truly HIV positive are 150 out of 151. It's a highly predictive test.

Let's now use the same test where the background HIV infection rate in the population is about one in ten thousand. If we test ten thousand people, we can expect two positive blood results overall: one from the person who really is HIV positive; and the one false positive that we could expect, again, from having ten thousand HIV negative men being tested with a test that is wrong one time in ten thousand.

Suddenly, when the background rate of an event is rare, even our previously brilliant blood test becomes a bit rubbish. For the

*The figures here are ballpark, from Gerd Gigerenzer's excellent book *Reckoning with Risk*.

two men with a positive HIV blood test result, in this population where only one in ten thousand has HIV, it's only fifty-fifty odds on whether they really are HIV positive.

Let's think about violence. The best predictive tool for psychiatric violence has a sensitivity of 0.75 and a specificity of 0.75. It's tougher to be accurate when we predict an event in humans, with human minds and changing human lives. Let's say 5 percent of patients seen by a community mental health team will be involved in a violent event in a year. If we use the same math as we did for the HIV tests, your "0.75" predictive tool would be wrong eighty-six times out of a hundred. For serious violence, occurring at 1 percent a year, with our best "0.75" tool, you inaccurately finger your potential perpetrator ninety-seven times out of a hundred. Will you preventively detain ninety-seven people to prevent three violent events? And will you apply that rule to alcoholics and assorted nasty antisocial types as well?

For murder, the extremely rare crime in question in this report, for which more action was demanded, occurring at one in ten thousand a year among patients with psychosis, the false positive rate is so high that the best predictive test is entirely useless.

This is not a counsel of despair. There are things that can be done, and you can always try to reduce the number of actual stark cock-ups, although it's difficult to know what proportion of the "one murder a week" represents a clear failure of a system, since when you look back in history, through the retrospectoscope, anything that happens will look as if it were inexorably leading up to your one bad event. I'm just giving you the math on rare events. What you do with it is a matter for you.

LOCKING YOU UP

In 1999 British lawyer Sally Clark was put on trial for murdering her two babies. In the U.K. this was a major trial, with a successful appeal, and although many have a dim awareness that there was a statistical error in the prosecution case, few know the true story or the phenomenal extent of the statistical ignorance that went on in the case.

At her trial, Professor Sir Roy Meadow, an expert in parents who harm their children, was called to give expert evidence. Meadow famously quoted "one in seventy-three million" as the chance of two children in the same family dying of sudden infant death syndrome (SIDS).

This was a very problematic piece of evidence for two very distinct reasons: one is easy to understand; the other is an absolute mind bender. Because you have the concentration span to follow the next two pages, you will come out smarter than Professor Sir Roy, the judge in the Sally Clark case, her defense teams, the appeal court judges, and almost all the journalists and legal commentators reporting on the case. We'll do the easy reason first.

THE ECOLOGICAL FALLACY

The figure of "one in seventy-three million" itself is iffy, as everyone now accepts. It was calculated as $8{,}543 \times 8{,}543$, as if the chances of two SIDS episodes in this one family were independent of each other. This feels wrong from the outset, and anyone can see why: there might be environmental or genetic factors at play, both of which would be shared by the two babies. But forget how pleased you are with yourself for understanding that fact. Even if we accept that two SIDS in one family is much more likely than one in seventy-three million—say, one in ten thousand—any such figure is still of dubious relevance, as we shall now see.

THE PROSECUTOR'S FALLACY

The real question in this case is: What do we do with this spurious number? Many press reports at the time stated that one in seventy-three million was the likelihood that the deaths of Sally Clark's two children were accidental—that is, the likelihood that she was innocent. Many in the court process seemed to share this view, and the factoid certainly sticks in the mind. But this is an example of a well-known and well-documented piece of flawed reasoning known as the prosecutor's fallacy.

Two babies in one family have died. This in itself is very rare. Once this rare event has occurred, the jury needs to weigh up two competing explanations for the babies' deaths: double SIDS or double murder. Under normal circumstances—before any babies have died—double SIDS is very unlikely, and so is double murder. But now that the rare event of two babies dying in one family has occurred, the two explanations—double murder or double SIDS—are suddenly both very likely. If we really wanted to play statistics, we would need to know which is relatively *more* rare: double SIDS or double murder. People have tried to calculate the relative risks of these two events, and one paper says it comes out at around two to one in favor of double SIDS.

Not only was this *crucial* nuance of the prosecutor's fallacy missed at the time—by everyone in the court—but it was also clearly missed in the appeal, at which the judges suggested that instead of "one in seventy-three million," Meadow should have said "very rare." They recognized the flaws in its calculation, the ecological fallacy, the easy problem above, but they still accepted his number as establishing "a very broad point, namely the rarity of double SIDS."

That, as you now understand, was entirely wrongheaded; the rarity of double SIDS is irrelevant, because double murder is rare too. An entire court process failed to spot the nuance of how the figure should be used. Twice.

Meadow was foolish, and has been vilified (some might say this process was exacerbated by the witch hunt against pediatricians who work on child abuse), but if it is true that he should have spotted and anticipated the problems in the interpretation of his number, then so should the rest of the people involved in the case: a pediatrician has no more unique responsibility to be numerate than a lawyer, a judge, journalist, jury member, or clerk. The prosecutor's fallacy is also highly relevant in DNA evidence, for example, in which interpretation frequently turns on complex mathematical and contextual issues. Anyone who is going to trade in numbers, and use them, and think with them, and persuade with them, let alone lock people up with them, also has a responsibility to understand them. All you've done is read a popular science book on them, and already you can see it's hardly rocket science.

LOSING THE LOTTERY

> You know, the most amazing thing happened to me tonight. I was coming here, on the way to the lecture, and I came in through the parking lot. And you won't believe what happened. I saw a car with the license plate ARW 357. Can you imagine? Of all the millions of license plates in the state, what was the chance that I would see that particular one tonight? Amazing . . .   —Richard Feynman

It is possible to be very unlucky indeed. A nurse named Lucia de Berk has been in prison for six years in Holland, convicted of seven counts of murder and three of attempted murder. An unusually large number of people died when she was on shift, and that, essentially, along with some very weak circumstantial evidence, is the substance of the case against her. She has never confessed, she has continued to protest her innocence, and her trial has generated a small collection of theoretical papers in the statistics literature.

The judgment was largely based on a figure of "one in 342 million against." Even if we found errors in this figure—and believe me, we will—as in our previous story, the figure itself would still be largely irrelevant. Because, as we have already seen repeatedly, the interesting thing about statistics is not the tricky math, but what the numbers mean.

There is also an important lesson here from which we could all benefit: unlikely things do happen. Somebody wins the lottery every week; children are struck by lightning. It's only weird and startling when something very, very specific and unlikely happens if you have specifically predicted it beforehand.*

Here is an analogy.

Imagine I am standing near a large wooden barn with an enormous machine gun. I place a blindfold over my eyes, and laughing maniacally, I fire off many thousands and thousands of bullets into the side of the barn. I then drop the gun, walk over to the wall, examine it closely for some time, all over, pacing up and down. I find one spot where there are three bullet holes close to one another, then draw a target around them, announcing proudly that I am an excellent marksman.

You would, I think, disagree with both my methods and my conclusions for that deduction. But this is exactly what has happened in Lucia's case: the prosecutors found seven deaths on one nurse's shifts, in one hospital, in one city, in one country, in the world and then drew a target around them.

This breaks a cardinal rule of any research involving statistics: you cannot find your hypothesis in your results. Before you go to your data with your statistical tool, you have to have a specific hypothesis to test. If your hypothesis comes from analyzing the data, then there is no sense in analyzing the same data again to confirm it.

This is a rather complex, philosophical, mathematical form of circularity, but there were also very concrete forms of circular rea-

---

*The magician and pseudoscience debunker James Randi used to wake up every morning and write on a card in his pocket: "I, James Randi, will die today," followed by the date and his signature. Just in case, he has recently explained, he really did, by some completely unpredictable accident.

soning in the case. To collect more data, the investigators went back to the wards to see if they could find more suspicious deaths. But all the people who were asked to remember "suspicious incidents" knew that they were being asked because Lucia might be a serial killer. There was a high risk that "an incident was suspicious" became synonymous with "Lucia was present." Some sudden deaths when Lucia was not present would not be listed in the calculations, by definition: they are in no way suspicious, because Lucia was not present.

It gets worse. "We were asked to make a list of incidents that happened during or shortly after Lucia's shifts," said one hospital employee. In this manner more patterns were unearthed, and so it became even more likely that investigators would find more suspicious deaths on Lucia's shifts. Meanwhile, Lucia waited in prison for her trial.

[This is the stuff of nightmares.]

At the same time, a huge amount of corollary statistical information was almost completely ignored. In the three years before Lucia worked on the ward in question, there were seven deaths. In the three years that she did work on the ward, there were six deaths. Here's a thought: it seems odd that the death rate should go *down* on a ward at the precise moment that a serial killer—on a killing spree—arrives. If Lucia killed them all, then there must have been no natural deaths on that ward at all in the whole of the three years that she worked there.

Ah, but on the other hand, as the prosecution revealed at her trial, Lucia did like tarot. And she does sound a bit weird in her private diary, excerpts from which were read out. So she might have done it anyway.

But the strangest thing of all is this. In generating his obligatory, spurious, Meadowesque figure, which this time was "one in 342 million," the prosecution's statistician made a simple, rudimentary mathematical error. He combined individual statistical tests by multiplying p-values, the mathematical description of chance, or statistical significance. This bit's for the hard-core science nerds, and will be edited out by the publisher, but I intend to write it anyway: you do not just multiply p-values together; you weave them with a clever tool, like maybe "Fisher's method for combination of independent p-values."

If you multiply p-values together, then harmless and probable incidents rapidly appear vanishingly unlikely. Let's say you worked in twenty hospitals, each with a harmless incident pattern, say, p=0.5. If you multiply those harmless p-values, of entirely chance findings, you end up with a final p-value of 0.5 to the power of twenty, which is $p < 0.000001$, which is extremely, very, highly statistically significant. With this mathematical error, by his reasoning, if you change hospitals a lot, you automatically become a suspect. Have you worked in twenty hospitals? For God's sake, don't tell the Dutch police if you have.

12

THE MEDIA'S MMR HOAX

In the previous chapter we looked at individual cases. They may have been egregious, and in some respects absurd, but the scope of the harm they can do is limited. We have already seen, with the example of Dr. Spock's advice to parents on how their babies should sleep, that when your advice is followed by a very large number of people, if you are wrong, even with the best of intentions, you can do a great deal of harm: because the effects of modest tweaks in risk are magnified by the size of the population changing its behavior.

It's for this reason that journalists have a special responsibility, and that's also why we will devote the last chapter of this book to examining the processes behind one very illustrative scare story: the MMR vaccine. But as ever, as you know, we are talking about much more than just that single tale, and there will be many distractions along the way.

In the United States, you are currently having your own vaccine scare. Jim Carrey and friends appear regularly on TV and in the newspapers to tell the nation of their concerns on complex matters of epidemiology and immunology. Foreigners—let me tell you a secret—like to sneer at Americans sometimes, for their crude popular debate. This is not justified. In the U.K., our vaccine scare was epic. It is coming to a close, but in the hope that you can learn something from it (and because one of its leading figures, Dr. Andrew Wakefield, has now moved to Texas and become your problem) here is the abysmal tale of MMR, the prototypical health scare, by which all others must be judged and understood.

Even now, it is with great trepidation that I dare mention it by name, because at the quietest hint of a discussion on the subject, an army of campaigners and columnists will still, even today, hammer on editors' doors, demanding the right to a lengthy, misleading, and emotive response in the name of "balance," in a world where their demands are always, without exception, accommodated.

At the beginning of this story, way back in 1998, is a man named Andrew Wakefield, who wrote a paper that linked the MMR vaccine to autism and set off a wave of antivaccine sentiment. This very month, as I write this chapter, the General Medical Council in the U.K. has found, after a two-year hearing, that he was "misleading," "dishonest," and "irresponsible" in the way he described where the children in his 1998 came from, by implying that they were routine clinic referrals. As the GMC also found, these children were subjected to a program of unpleasant and invasive tests that were performed not in their own clinical interest, but rather for research purposes, and these tests were conducted without ethics committee approval. It's plainly undesirable for doctors to go around conducting tests like colonoscopy on children for their own interest.

But as we shall see, Dr. Wakefield cannot carry the blame for this scare alone, however much the news media may now try to imply that he should; the blame lies instead with the hundreds of journalists, columnists, editors, and executives, in every single news outlet in the U.K., who drove this story cynically, irrationally, and willfully onto the front pages for nine solid years. As we shall also see, they overextrapolated from one study into absurdity, while studiously ignoring all reassuring data and all subsequent refutations. They quoted "experts" as authorities instead of explaining the science, they ignored the historical context, they set idiots to cover the facts, they pitched emotive stories from parents against bland academics (whom they smeared), and most bizarrely of all, in some cases they simply made stuff up.

Journalists frequently flatter themselves with the fantasy that they are unveiling vast conspiracies, that the entire medical establishment has joined hands to suppress an awful truth. In reality I would guess that the 150,000 doctors in the U.K. could barely agree on second-line management of hypertension, but no matter: this fantasy was the structure of the MMR story, and many others, but it was a similar grandiosity that drove many of the earlier examples in this book in which a journalist concluded that he knew best, including "Cocaine use doubles in the playground."

In some respects, this reflects changes in the environment for investigative journalism; this kind of work is expensive and risks expensive legal cases from the powerful people you investigate. Concocting a health scare is attractive, because it gives the appearance of challenging power and authority, but with none of the work, and none of the litigation risk if you're wrong.

But can they ever do good? Undoubtedly there must be some examples, but the imperfect systems of medicine catch errors with far greater frequency. Often, to my surprise, journalists will cite "thalidomide" as if this were investigative journalism's greatest triumph in medicine, in which they bravely exposed the risks of the drug in the face of medical indifference. It comes up almost every time I lecture on the media's crimes in science, and that is why I will explain the story in some detail here, because in reality— sadly, really—this finest hour never occurred.

In 1957, a baby was born with no ears to the wife of an employee at Grünenthal, the German drug company. He had taken its new antinausea drug home for his wife to try while she was pregnant, a full year before it went on the market. This is an illustration both of how slapdash things were and of how difficult it is to spot a pattern from a single event.

The drug went to market, and between 1958 and 1962 around ten thousand children all around the world were born with severe malformations, caused by this same drug, thalidomide. Because there was no central monitoring of malformations or adverse reactions, the pattern was missed. An Australian obstetrician called William McBride first raised the alarm in a medical journal, publishing a letter in *The Lancet* in December 1961. He ran a large obstetric unit, seeing a great number of cases, and he was rightly regarded as a hero; but it's sobering to think that he was in such a good position to spot the pattern only because he had prescribed so much of the drug, without knowing its risks, to his patients.* By the time his letter was published, a German pediatrician had noted a similar pattern, and the results of his study had been described in a German Sunday newspaper a few weeks earlier.

Almost immediately afterward, the drug was taken off the market, and pharmacovigilance began in earnest, with notification schemes set up around the world, however imperfect you may find them to be. If you ever suspect that you've experienced an adverse drug reaction, I would regard it as your duty as a member of the public, to report it (in the United States anyone, including patients, can report an adverse event at the FDA MedWatch site). These reports can be collated and monitored as an early warning sign, and are a part of the imperfect, pragmatic monitoring system for picking up problems with medications.

Now they claim that the original 1998 Wakefield research has been "debunked" (it was never anything compelling in the first place), and you will be able to watch this year as they try to pin the whole scare onto one man. I'm a doctor too, and I don't imag-

*Many years later William McBride turned out to be guilty, in an unfortunate twist, of research fraud, falsifying data, and he was struck off the medical register in 1993. He was later reinstated.

ine for one moment that I could stand up and create a nine-year-long news story on a whim. It is because of the media's blindness—and their unwillingness to accept their responsibility—that they will continue to commit the same crimes in the future. There is nothing you can do about that, so it might be worth paying attention now.

To remind ourselves, here is the story of MMR as it appeared in the British news media from 1998 onward:

- Autism is becoming more common, although nobody knows why.
- A doctor called Andrew Wakefield has done scientific research showing a link between the MMR triple jab and autism.
- Since then, more scientific research has been done confirming this link.
- There is evidence that single jabs might be safer, but government doctors and those in the pay of the pharmaceutical industry have simply rubbished these claims.
- Tony Blair probably didn't give his young son the vaccine.
- Measles isn't so bad.
- And vaccination didn't prevent it very well anyway.

I think that's pretty fair. The central claim for each of these bullet points was either misleading or downright untrue, as we shall see.

VACCINE SCARES IN CONTEXT

Before we begin, it's worth taking a moment to look at vaccine scares around the world, because I'm always struck by how circumscribed these panics are and how poorly they propagate themselves in different soils. Before celebrities got their hands on it, a decade later, the MMR and autism scare, for example, was practically nonexistent outside Britain, even in Europe and the United States. But throughout the 1990s France was in the grip of a scare that hepatitis B vaccine caused multiple sclerosis (it wouldn't surprise me if I were the first person to tell you that).

In the United States, at the time, the major vaccine fear had been around the use of a preservative called thimerosal, although somehow this hasn't caught on in the U.K., even though that same preservative was used in Britain. And in the 1970s—since the past is another country too—there was a widespread concern in the U.K., driven again by a single doctor, that whooping cough vaccine was causing neurological damage.

To look even farther back, there was a strong anti–smallpox vaccine movement in Leicester well into the 1930s, despite its demonstrable benefits, and in fact, anti-inoculation sentiment goes right back to its origins: when James Jurin studied inoculation against smallpox (finding that it was associated with a lower death rate than the natural disease), his newfangled numbers and statistical ideas were treated with enormous suspicion. Indeed smallpox inoculation remained illegal in France until 1769.* Even when Edward Jenner introduced the much safer vaccination for protecting people against smallpox at the turn of the nineteenth century, he was strongly opposed by the London cognoscenti.

---

*Disdain for statistics in health care research wasn't unusual at the time: Ignaz Semmelweis noticed in 1847 that patients were dying much more frequently on the obstetrics ward run by the medical students than on the one run by the midwifery students (this was in the days when students did all the legwork in hospitals). He was pretty sure that this was because the medical students were carrying something nasty from the corpses in the dissection room, so he instituted proper hand-washing practices with chlorinated lime and did some figures on the benefits. The death rates fell, but in an era of medicine that championed "theory" over real-world empirical evidence, he was basically ignored, until Louis Pasteur came along and confirmed the germ theory. Semmelweis died alone in an asylum. You've heard of Pasteur.

And in an article from *Scientific American* in 1888 you can find the very same arguments that modern antivaccination campaigners continue to use today:

> The success of the anti-vaccinationists has been aptly shown by the results in Zurich, Switzerland, where for a number of years, until 1883, a compulsory vaccination law obtained, and smallpox was wholly prevented—not a single case occurred in 1882. This result was seized upon the following year by the anti-vaccinationists and used against the necessity for any such law, and it seems they had sufficient influence to cause its repeal. The death returns for that year (1883) showed that for every 1,000 deaths two were caused by smallpox; In 1884 there were three; in 1885, 17, and in the first quarter of 1886, 85.

Meanwhile, WHO's highly successful global polio eradication program was on target to have eradicated this murderous disease from the face of the earth by now—a fate that has already befallen the smallpox virus, excepting a few glass vials—until local imams from a small province called Kano in northern Nigeria claimed that the vaccine was part of a U.S. plot to spread AIDS and infertility in the Islamic world and organized a boycott that rapidly spread to five other states in the country. This was followed by a large outbreak of polio in Nigeria and surrounding countries and tragically even farther afield. There have now been outbreaks in Yemen and Indonesia, causing lifelong paralysis in children, and laboratory analysis of the genetic code has shown that these outbreaks were caused by the same strain of the polio virus, exported from Kano.

After all, as any trendy MMR-dodging North London middle-class humanities graduate couple with children would agree, just because vaccination has almost eradicated polio—a debilitating disease that as recently as 1988 was endemic in 125 countries—doesn't necessarily mean it's a good thing.

The diversity and isolation of these antivaccination panics help illustrate the way in which they reflect local political and social concerns more than a genuine appraisal of the risk data: because if the vaccine for hepatitis B, or MMR, or polio is dangerous in one country, it should be equally dangerous everywhere on the planet, and if those concerns were genuinely grounded in the evidence, especially in an age of the rapid propagation of information, you would expect the concerns to be expressed by journalists everywhere. They're not.

ANDREW WAKEFIELD AND
HIS *LANCET* PAPER

In February 1998 a group of researchers and doctors led by a surgeon called Andrew Wakefield from the Royal Free Hospital in London published a research paper in *The Lancet* that by now stands as one of the most misunderstood and misreported papers in the history of academia. In some respects it did itself no favors: it is badly written and has no clear statement of its hypothesis, or indeed of its conclusions (you can read it free online if you like). It has since been fully retracted by *The Lancet*, whose editor explained it was "utterly clear, without any ambiguity at all, that the statements in the paper were utterly false."

The paper described twelve children who had bowel problems and behavioral problems (mostly autism) and mentioned that the parents or doctors of eight of these children believed that their children's problems had started within a few days of their being given the MMR vaccine. It also reported various blood tests and tests on tissue samples taken from the children. The results of these were sometimes abnormal, but varied between children.

12 children, consecutively referred to the department of pediatric gastroenterology with a history of a pervasive developmental disorder with loss of acquired skills and intestinal symptoms (diarrhea, abdominal pain, bloating and food intolerance), were investigated.

. . . In eight children, the onset of behavioral problems had been linked, either by the parents or by the child's physician, with measles, mumps, and rubella vaccination . . . In these eight children the average interval from exposure to first behavioral symptoms was 6.3 days (range 1–14).

What can this kind of paper tell you about a link between something as common as MMR and something as common as autism? Basically nothing, either way. It was a collection of twelve clinical anecdotes, a type of paper called a case series, and a case series, by design, wouldn't demonstrate such a relationship between an exposure and an outcome with any force. It did not take some children who were given MMR and some children who weren't and then compare the rates of autism between the two groups (this would have been a cohort study). It did not take some children with autism and some children without autism and then compare the rates of vaccination between the two groups (this would have been a case-control study).

Could anything else explain the apparent connection among MMR, bowel problems, and autism in these eight children? First, although they sound like rare things to come together, this was a specialist center in a teaching hospital, and the children had been referred there only because they had bowel problems and behavioral problems (the circumstances of these referrals are currently being examined by the GMC, as we shall see).

Out of an entire nation of millions of inhabitants, if some children with a combination of fairly common things (vaccination, autism, bowel problems) all come together in one place that is *al-*

ready acting as a beacon for such a combination, as this clinic was, we should not naturally be impressed. You will remember from the discussion of the unfortunate Dutch nurse Lucia de Berk (and indeed from reading news reports about lottery winners) that unlikely combinations of events will always happen, somewhere, to some people, entirely by chance. Drawing a target around them after the fact tells us nothing at all.

All stories about treatment and risk will start with modest clinical hunches like these anecdotes: but hunches, with nothing to back them up, are not generally newsworthy. At the publication of this paper, a press conference was held at the Royal Free Hospital, and to the visible surprise of many other clinicians and academics present, Andrew Wakefield announced that he thought it would be prudent to use single vaccines instead of the MMR triple vaccine. Nobody should have been surprised; a video news release had already been issued by the hospital in which Wakefield made the same call.

We all are entitled to our clinical hunches as individuals, but there was nothing in either this study of twelve children or any other published research to suggest that giving single vaccines would be safer. As it happens, there are good grounds for believing that giving vaccines separately might be more harmful: they need six visits to the GP, and six unpleasant jabs, making four more appointments to miss. Maybe you're ill, maybe you're on holiday, maybe you move house, maybe you lose track of which ones you've had, maybe you can't see the point of rubella for boys, or mumps for girls, or maybe you're a working single mum with two kids and no time.

Also, of course, the children spend much more time vulnerable to infection, especially if you wait a year between jabs, as Wakefield has recommended, out of the blue. Ironically, although most of the causes of autism remain unclear, one of the few well-characterized single causes is rubella infection itself while the child is in the womb.

THE STORY BEHIND THE PAPER

Since then this paper has been entirely discredited. I don't want that aspect of the story—rather than the research evidence—to be the reason why you come to your own conclusion about the risks of MMR and autism. There are things that came out in 2004, however, that cannot fairly be ignored; they include allegations of multiple conflicts of interest, undeclared sources of bias in the recruitment of subjects for the paper, undisclosed negative findings, and problems with the ethical clearance for the tests. These were largely uncovered by a tenacious investigative journalist from *The Sunday Times* called Brian Deer, and they formed part of a case brought against Andrew Wakefield by the GMC, the medical regulator in the U.K.

While in the paper it is stated that the children investigated were sequential referrals to a clinic, in fact, Wakefield was already being paid seventy-five thousand dollars of legal aid money by a firm of solicitors to investigate children whose parents were preparing a case against MMR, and many of these referrals had come to him specifically as someone who could show a link between MMR and autism, whether formally or informally, and was working on a legal case. This is the beacon problem once more, and under these circumstances, the fact that *only* eight of the twelve children's parents or physicians believed the problems were caused by MMR would be unimpressive, if anything.

Of the twelve children in the paper, eleven sued drug companies, and ten of them already had legal aid to sue over MMR before the 1998 paper was published. Wakefield himself eventually received almost $700,000 plus expenses from the legal aid fund for his role in the case against MMR.

The GMC has found that various intrusive clinical investigations, such as lumbar punctures and colonoscopies, were carried out on the children, not to determine their own treatment but rather for research purposes; furthermore, these tests were conducted without ethics committee approval.

Lumbar puncture involves putting a needle into the center of the spine to tap off some spinal fluid, and colonoscopy involves putting a flexible camera and light through the anus, up the rectum, and into the bowel on a long tube. Neither is without risk, and indeed one of the children being investigated as part of an extension of the MMR research project was seriously harmed during colonoscopy and was rushed to intensive care at Great Ormond Street Hospital after his bowel had been punctured in twelve places. He suffered multiple organ failure, including kidney and liver problems, and neurological injuries, and received $740,000 in compensation. These things happen, nobody is to blame, and I am merely illustrating the reasons to be cautious about doing investigations.

Outside of this, there are also other issues, uncovered by Brian Deer. In 1997 a young Ph.D. student called Nick Chadwick was starting his research career in Andrew Wakefield's lab, using PCR technology (used as part of DNA fingerprinting) to look for traces of measles strain genetic material in the bowels of these twelve children, because this was a central feature of Wakefield's theory. In 2004 Chadwick gave an interview to Channel 4's *Dispatches*, and in 2007 he gave evidence at a U.S. case on vaccines, stating that there was no measles RNA to be found in these samples. But this important finding, which conflicted with his charismatic supervisor's theory, was not published.

I could go on.

Nobody knew about any of this in 1998. In any case, it's not relevant, because the greatest tragedy of the media's MMR hoax is that it was brought to an end by these issues' being made public, when it should have been terminated by a cautious and balanced appraisal of the evidence at the time. Now, you will see news reporters, including the BBC, saying stupid things like "The research has since been debunked." Wrong. The research never justified the media's ludicrous overinterpretation. If they had paid attention, the scare would never have even started.

THE PRESS COVERAGE BEGINS

What's most striking about the MMR scare—and this is often forgotten—is that it didn't actually begin in 1998. *The Guardian* and *The Independent* covered the press conference on their front pages, but *The Sun* ignored it entirely, and the *Daily Mail*, international journal of health scares, buried its piece on it in the middle of the paper. Coverage of the story was generally written by specialist health and science journalists, and they were often fairly capable of balancing the risks and evidence. The story was pretty soft.

In 2001 the scare began to gain momentum. Wakefield published a review paper in an obscure journal, questioning the safety of the immunization program, although with no new evidence. In March he published new laboratory work with Japanese researchers ("the Kawashima paper"), using PCR data to show measles virus in the white blood cells of children with bowel problems and autism. This was essentially the opposite of the findings from Nick Chadwick in Wakefield's own lab. Chadwick's work remained unmentioned (and there has since been a paper published showing how the Kawashima paper produced a false positive, although the media completely ignored this development, and Wakefield seems to have withdrawn his support for the study).

Things began to deteriorate. The antivaccination campaigners began to roll their formidable and coordinated publicity machine into action against a rather chaotic shambles of independent doctors from various different uncoordinated agencies. Emotive anecdotes from distressed parents were pitted against old duffers in corduroy, with no media training, talking about scientific data. If you ever wanted to see evidence against the existence of a sinister medical conspiracy, you need look no further than the shower of avoidant doctors and academics and their piecemeal engagement with the media during this time. The Royal College of General Practitioners not only failed to speak clearly on the evidence. They also managed—heroically—to dig up some anti-MMR GPs to offer to journalists when they rang in asking for quotes.

The story, perhaps bound up in the wider desire of some newspapers and personalities simply to attack the government and the health service, began to gain momentum. A stance on MMR became part of many newspapers' editorial policies, and that stance was often bound up with rumors about senior managerial figures with family members who had been affected by autism. It was the perfect story, with a single charismatic maverick fighting against the system, a Galileo-like figure; there were elements of risk, of awful personal tragedy, and, of course, the question of blame. Whose fault was autism? Because nestling in the background was this extraordinary new diagnosis, a disease that struck down young boys and seemed to have come out of the blue, without explanation.

AUTISM

We still don't know what causes autism. A history of psychiatric problems in the family, early birth, problems at birth, and breech presentation all are risk factors, but pretty modest ones, which means they're interesting from a research perspective, but none of them explains the condition in a particular person. This is often the case with risk factors. Boys are affected more than girls, and the incidence of autism continues to rise, in part because of improved diagnosis—people who were previously given labels like "mentally subnormal" or "schizophrenia" were now receiving a diagnosis of "autism"—but also possibly because of other factors that are still not understood. Into this vacuum of uncertainty, the MMR story appeared.

There was also something strangely attractive about autism as an idea to journalists and other commentators. Among other things, it's a disorder of language, which might touch a particular chord with writers; but it's also philosophically enjoyable to think about, because the flaws in social reasoning that are exhibited by people with autism give us an excuse to talk and think about our social norms and conventions. Books about autism and the autistic outlook on the world have become bestsellers. Here are some wise words for us all from Luke Jackson, a thirteen-year-old with Asperger's syndrome, who has written a book of advice for teenagers with the condition (*Freaks, Geeks and Asperger Syndrome*). This is from the section on dating:

> If the person asks something like "Does my bum look fat?" or even "I am not sure I like this dress" then that is called "fishing for compliments." These are very hard things to understand, but I am told that instead of being completely honest and saying that yes their bum does look fat, it is politer to answer with something like "Don't be daft, you look great." You are not lying, simply evading an awkward question and complimenting them at the same time. Be economical with the truth!

Asperger's syndrome, or autistic spectrum disorder, is being applied to an increasingly large number of people, and children or adults who might previously have been considered "quirky" now frequently have their personalities medicalized with suggestions that they have "traits of Asperger's." Its growth as a pseudodiagnostic category has taken on similar proportions to "mild dyslexia"—you will have your own views on whether this process is helpful—and its widespread use has allowed us all to feel that we can participate in the wonder and mystery of autism, each with a personal connection to the MMR scare.

Except of course, in most cases, genuine autism is a pervasive developmental disorder, and most people with autism don't write quirky books about their odd take on the world that reveal so much to us about our conventions and social mores in a charmingly plain and unself-conscious narrative style. Similarly, most people with autism do not have the telegenic single skills that the media have so enjoyed talking up in their crass documentaries, like being *really amazing* at mental arithmetic or playing the piano to concert standard while staring confusedly into the middle distance.

That these are the sorts of things most people think of when the word "autism" pops into their heads is testament to the mythologization and paradoxical "popularity" of the diagnosis. Mike Fitzpatrick, a physician with a son who has autism, says that there are two questions on the subject that will make him want to slap you. One is: "Do you think it was caused by MMR?" The other is: "Does he have any special skills?"

LEO BLAIR

But the biggest public health disaster of all was a sweet little baby called Leo. In December 2001 Prime Minister Tony Blair and his wife, Cherie, were asked if their infant son had been given the MMR vaccine and refused to answer. Most other politicians have been happy to clarify whether their children have had the vaccine, but you can imagine how people might believe the Blairs were the kind of family not to have their children immunized, especially with everyone talking about "herd immunity" and the worry that they might be immunizing their child, and placing it at risk, in order that the rest of the population should be safer.

Concerns were particularly raised by the ubiquity of Cherie Blair's closest friend and aide. Carole Caplin was a New Age guru, a "life coach," and a "people person," although her boyfriend, Peter Foster, was a convicted fraudster. Foster helped arrange the Blairs' property deals, and he also says that they took Leo to a New Age healer, Jack Temple, who offered crystal dowsing, homoeopathy, herbalism, and Neolithic-circle healing in his back garden.

I'm not sure how much credence to give to Foster's claims myself, but the impact on the MMR scare is that they were widely reported at the time. We were told that the prime minister of the United Kingdom agreed to Temple's waving a crystal pendulum over his son to protect him (and therefore his classmates, of course) from measles, mumps, and rubella and that Tony let Cherie give Temple some of his own hair and nail clippings, which Temple preserved in jars of alcohol. He said he only needed to swing his pendulum over the jar to know if their owner was healthy or ill.

Some things are certainly true. Using this crystal dowsing pendulum, Temple did claim that he could harness energy from heavenly bodies. He sold remedies with names like Volcanic Memory, Rancid Butter, Monkey Sticks, Banana Stem, and, my own personal favorite, Sphincter. He was also a very well-connected man. Jerry Hall endorsed him. The duchess of York wrote the introduction to his book *The Healer: The Extraordinary Healing Methods of Jack Temple* (it's a hoot). He told the *Daily Mail* that babies who are breastfed from the moment of birth acquire natural immunity against all diseases, and he even sold a homoeopathic alternative to the MMR jab.

"I tell all my patients who are pregnant that when the baby is born they must put it on the breast until there is no longer a pulse in the umbilical cord. It usually takes about 30 minutes. By doing this they transfer the mother's immune system to the baby, who will then have a fully-functioning immune system and will not need vaccines." ... Mr Temple refused to confirm yesterday whether he advised Mrs Blair not to have her baby Leo vaccinated. But he said: "If women follow my advice their children will not need the MMR injection, end of story."*

—Daily Mail, *December 26, 2001*

Cherie Blair was also a regular visitor to Carole's mum, Sylvia Caplin, a spiritual guru. "There was a particularly active period in the summer when Sylvia was channelling for Cherie over two or three times a week, with almost daily contact between them," the *Mail* reported. "There were times when Cherie's faxes ran to 10 pages." Sylvia, along with many, if not most, alternative therapists, was viciously anti-MMR (over half of all the homeopaths approached in one survey grandly advised against the vaccine). *The Daily Telegraph* reported:

We move on to what is potentially a very political subject: the MMR vaccine. The Blairs publicly endorsed it, then caused a minor furore by refusing to say whether their baby, Leo, had been inoculated. Sylvia [Caplin] doesn't hesitate: "I'm against it," she says. "I'm appalled at so much being given to little children. The thing about these drugs is the toxic substance they put the vaccines in—for a tiny child, the MMR is a ridiculous thing to do.

"It has definitely caused autism." All the denials that Fuck Off come from the old school of medicine are open to ques-

*Here is Jack on cramp: "For years many people have suffered with cramp. By dowsing, I discovered that this is due to the fact that the body is not absorbing the element 'scandium' which is linked to and controls the absorption of magnesium phosphate." And on general health complaints: "Based on my expertise in dowsing, I noted that many of my patients were suffering from severe deficiencies of carbon in their systems. The ease in which people these days suffer hairline fractures and broken bones is glaringly apparent to the eyes that are trained to see."

tion because logic and common sense must tell you that there's some toxic substance in it. Do you not think that's going to have an effect on a tiny child? Would you allow it? No—too much, too soon, in the wrong formula."

It was also reported—doubtless as part of a cheap smear—that Cherie Blair and Carole Caplin encouraged the prime minister to have Sylvia "douse and consult The Light, believed by Sylvia to be a higher being or God, by use of her pendulum" to decide if it was safe to go to war in Iraq. And while we're on the subject, in December 2001 *The Times* (London) described the Blairs' holiday in Temazcal, Mexico, where they rubbed fruits and mud over each other's bodies inside a large pyramid on the beach, then screamed while going through a New Age rebirthing ritual. Then they made a wish for world peace.

I'm not saying I buy all this. I'm just saying, this is what people were thinking about when the Blairs refused to clarify publicly the issue of whether they had given their child the MMR vaccine as all hell broke loose around it. This is not a hunch. Of all the stories written that year about MMR, 32 percent mentioned whether Leo Blair had had the vaccine or not (even Andrew Wakefield was only mentioned in 25 percent), and it was one of the most well-recalled facts about the story in population surveys. The public, quite understandably, was taking Leo Blair's treatment as a yardstick of the prime minister's confidence in the vaccine, and few could understand why it should be a secret if it wasn't an issue.

The Blairs, meanwhile, cited their child's right to privacy, which they felt was more important than an emerging public health crisis. It's striking that Cherie Blair has now decided, in marketing her lucrative autobiography, to waive that principle, which was so vital at the time, and has written at length in her heavily promoted book not just about the precise bonk that conceived Leo but also about whether he had the jab (she says yes, but she seems to obfuscate on whether it was single vaccines and indeed on the question of when he had it; frankly, I give up on these people).

For all that it may seem trite and voyeuristic to you, this event was central to the UK coverage of MMR. The year 2002 was the year of Leo Blair, and of Wakefield's departure from the Royal Free, and it was the peak of the media coverage, by a very long margin.

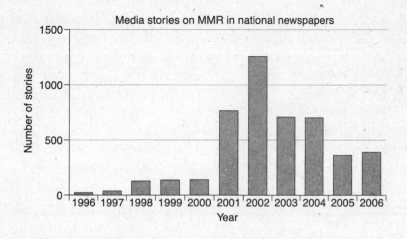

WHAT WAS IN THESE STORIES?

The MMR scare has created a small cottage industry of media analysis, so there is a fair amount known about the coverage. In 2003 the Economic and Social Research Council (ESRC) published a paper on the media's role in the public understanding of science, which sampled all the major science media stories from January to September 2002, the peak of the scare. Ten percent of all science stories were about MMR, and MMR was also by far the most likely to generate letters to the press (so people were clearly engaging with the issue), by far the most likely science topic to be written about in opinion or editorial pieces, and it generated the longest stories. MMR was the biggest, most heavily covered science story for years.

Pieces on genetically modified (GM) food or cloning stood a good chance of being written by specialist science reporters, but for stories on MMR these reporters were largely sidelined, and 80 percent of the coverage of the biggest science story of the year was by generalist reporters. Suddenly we were getting comment and advice on complex matters of immunology and epidemiology from people who would more usually have been telling us about a funny thing that happened with the au pair on the way to a dinner party. Nigella Lawson, Libby Purves, Suzanne Moore, Lynda Lee-Potter, and Carol Vorderman, to name only a few, all wrote about their ill-informed concerns on MMR, blowing hard on their toy trumpets. The anti-MMR lobby, meanwhile, developed a reputation for targeting generalist journalists wherever possible, feeding them stories, and actively avoiding health or science correspondents.

This is a pattern that has been seen before. If there is one thing that has adversely affected communication among scientists, journalists, and the public, it is the fact that science journalists simply do not cover major science news stories. From drinking with science journalists, I know that much of the time, nobody even runs these major stories by them for a quick check.

Again, I'm not speaking in generalities here. During the crucial two days after the GM "Frankenstein foods" story broke in February 1999, *not a single one* of the news articles, opinion pieces, or editorials on the subject was written by a science journalist. A science correspondent would have told his or her editor that when someone presents his scientific findings about GM potatoes' causing cancer in rats, as Árpád Pusztai did, on ITV's *World in Action*, rather than in an academic journal, there's something fishy going on. Pusztai's experiment was finally published a year later—after a long period when nobody could comment on it, because nobody knew what he'd actually done—and when all was revealed in a proper publication, his experimental results did not contain information to justify the media's scare.

This sidelining of specialist correspondents when science becomes front-page news, and the fact that they are not even used as a resource during these periods, have predictable consequences. Journalists are used to listening with a critical ear to briefings from press officers, politicians, PR executives, salespeople, lobbyists, celebrities, and gossipmongers, and they generally display a healthy natural skepticism, but in the case of science, they don't have the skills to critically appraise a piece of scientific evidence on its merits. At best the evidence of these "experts" will be examined only in terms of who they are as people or perhaps whom they have worked for. Journalists—and many campaigners—think that this is what it means to critically appraise a scientific argument and seem rather proud of themselves when they do it.

The scientific content of stories—the actual experimental evidence—is brushed over and replaced with didactic statements from authority figures on either side of the debate, which contributes to a pervasive sense that scientific advice is somehow arbitrary and predicated upon a social role—the "expert"—rather than on transparent and readily understandable empirical evidence. Worse than this, other elements are brought into the foreground: political issues, Tony Blair's refusal to say whether his baby had received the vaccine, mythical narratives, a lionized "maverick" scientist, and emotive appeals from parents.

A reasonable member of the public, primed with such a compelling battery of human narrative, would be perfectly entitled to regard any expert who claimed MMR was safe as thoughtless and dismissive, especially if that claim came without any apparent supporting evidence.

The story was also compelling because, like GM food, the MMR story seemed to fit a fairly simple moral template, and one that I myself would subscribe to: big corporations are often dodgy, and politicians are not to be trusted. But it matters whether your political and moral hunches are carried in the right vehicle. Speaking only for myself, I am very wary of drug companies, not because I think all medicine is bad, but because I know they have hidden unflattering data and because I have seen their promotional material misrepresent science. I also happen to be very wary of GM food—but not because of any inherent flaws in the technology and not because I think it is uniquely dangerous. Somewhere between splicing in genes for products that will treat hemophilia at one end and releasing genes for antibiotic resistance into the wild at the other lies a sensible middle path for the regulation of GM, but there's nothing desperately remarkable or uniquely dangerous about it as a technology.

Despite all that, I remain extremely wary of GM for reasons that have nothing to do with the science, simply because it has created a dangerous power shift in agriculture, and "terminator seeds," which die at the end of the season, are a way to increase farmers' dependency, both nationally and in the developing world, while placing the global food supply in the hands of multinational corporations. If you really want to dig deeper, Monsanto is also very simply an unpleasant company (it made Agent Orange during the Vietnam War, for example).

Witnessing the blind, seething, thoughtless campaigns against MMR and GM, which mirror the infantile train of thought that "homeopathy works because the Vioxx side effects were covered up by Merck," it's easy to experience a pervasive sense of lost political opportunities, that somehow all our valuable indignation about development issues, the role of big money in our society, and frank corporate malpractice is being diverted away from anywhere it could be valid and useful and into puerile, mythical fantasies. It strikes me that if you genuinely care about big business, the environment, and health, then you're wasting your time with jokers like Pusztai and Wakefield.

Science coverage is further crippled, of course, by the fact that the subject can be quite difficult to understand. This in itself can seem like an insult to intelligent people, like journalists, who fancy themselves able to understand most things, but there has also been an acceleration in complexity in recent times. Fifty years ago you could sketch out a full explanation of how an AM radio worked on the back of a napkin, using basic school-level knowledge of science, and build a crystal set in a classroom that was essentially the same as the one in your car. When your parents were young, they could fix their own car and understand the science behind most of the everyday technology they encountered, but this is no longer the case. Even a geek today would struggle to give an explanation of how his mobile phone works, because technology has become more difficult to understand and explain, and everyday gadgets have taken on a "black box" complexity that can feel sinister, as well as intellectually undermining. The seeds were sown.

But we should return to the point. If there was little science, then what *did* appear in all these long stories on MMR? Going back to the 2002 data from the ESRC, only a quarter mentioned Andrew Wakefield, which seems odd, considering he was the cornerstone of the story. This created the erroneous impression that there was a large body of medical opinion that was suspicious of MMR, rather than just one "maverick." Less than a third of broadsheet reports referred to the overwhelming evidence that MMR is safe, and only 11 percent mentioned that it is regarded as safe in the ninety other countries in which it is used.

It was rare to find much discussion of the evidence at all, as it was considered too complicated, and when doctors tried to explain it, they were frequently shouted down, or worse still, their explanations were condensed into bland statements that "science had shown" there was nothing to worry about. This uninformative dismissal was pitted against the emotive concerns of distressed parents.

As 2002 wore on, things got really strange. Some newspapers made MMR the focus of a massive political campaign, and the beatification of Wakefield reached a kind of fever pitch. Lorraine Fraser had an exclusive interview with him in the *Telegraph* in which he was described as "a champion of patients who feel their fears have been ignored." She wrote a dozen similar articles over the next year (and her reward came when she was named British Press Awards Health Writer of the Year 2002, a gong I do not myself expect to receive).

Justine Picardie did a lavish photo feature on Wakefield, his house, and his family for the *Telegraph* Saturday magazine. Andy is, she tells us, "a handsome, glossy-haired hero to families of autistic children." How is the family? "A likeable, lively family, the kind you would be happy to have as friends, pitted against mysterious forces who have planted bugging devices and have stolen patients' records in 'apparently inexplicable' burglaries." She fantasizes— and I absolutely promise you I'm not making this up—about a Hollywood depiction of Wakefield's heroic struggle, with Russell Crowe playing the lead "opposite Julia Roberts as a feisty single mother fighting for justice for her child."

THE EVIDENCE ON MMR

So what is the evidence on the safety of MMR?

There are a number of ways to approach the evidence on the safety of a given intervention, depending on how much attention you have to give. The simplest approach is to pick an arbitrary authority figure: a doctor, perhaps, although this seems not to be appealing (in surveys people say they trust doctors the most and journalists the least; this shows the flaw in that kind of survey).

You could take another, larger authority at face value, if there is one that suits you. The Institute of Medicine, the Royal Colleges, the NHS, and more all came out in support of MMR, but this was apparently not sufficient to convince. You could offer information: an NHS website at mmrthefacts.nhs.uk started with the phrase "MMR is safe" (literally) and allowed the reader to drill down to the detail of individual studies.* But that too did little to stem the tide. Once a scare is running, perhaps every refutation can seem like an admission of guilt, drawing attention to the scare.

The Cochrane Collaboration is as blemishless as they come, and it has done a systematic review of the literature on MMR, concluding that there was no evidence that the vaccine is unsafe (although the story didn't appear until 2005). This reviewed the data the media had systematically ignored: What was in it?

If we are to maintain the moral high ground, there are a few things we need to understand about evidence. First, there is no single golden study that proves that MMR is safe (although the evidence to say it is dangerous was exceptionally poor). There is, for example, no randomized controlled trial. We are presented instead with a huge morass of data, from a number of different studies, all of which are flawed in their own idiosyncratic ways for reasons of cost, competence, and so on. A common problem with applying old data to new questions is that these papers and data sets might have a lot of useful information, which was collected very competently to answer the questions that the researchers were interested in at the time, but that isn't perfect for your needs. It's just, perhaps, pretty good.

Smeeth et al., for example, did something called a case-control study, using the GP Research Database. This is a common type of study, in which you take a bunch of people with the condition you're

---

*Whether you buy the Department of Health phrase "MMR is safe" depends on what you decide you mean by "safe." Is flying safe? Is your washing machine safe? What are you sitting on? Is that safe? You can obsess over the idea that philosophically nothing can ever be shown to be 100 percent safe, and many will, but you would be arguing about a fairly meaningless and uncommon definition of the word.

looking at ("autism") and a bunch of people without it, then look to
see if there is any difference in how much each group was exposed
to the thing you think might be causing the condition ("MMR").

If you care who paid for the study—and I hope you've become
a bit more sophisticated than that by now—it was funded by the
Medical Research Council. It found around thirteen hundred people with autism and then got some "controls," random people who
did not have autism, but with the same age, sex, and general practice. Then they looked to see if vaccination was any more common
in the people with autism or the controls and found no difference
between the two groups. The same researchers also did a systematic
review of similar studies in the United States and Scandinavia, and
again, pooling the data, found no link between MMR and autism.

There is a practical problem with this kind of research, of
course, that I would hope you might spot: most people *do* get the
MMR vaccine, so the individuals you're measuring who *didn't* get
the vaccine might be unusual in other ways—perhaps their parents have refused the vaccine for ideological or cultural reasons,
or the child has a preexisting physical health problem—and those
factors might themselves be related to autism. There's little you
can do in terms of study design about this potential "confounding
variable," because as we said, you're not likely to do a randomized
controlled trial in which you randomly don't give children vaccines;
you just throw the result into the pot with the rest of the information, in order to reach your verdict. As it happens, Smeeth et al.
went to great lengths to make sure their controls were representative. If you like, you can read the paper and decide if you agree.

So "Smeeth" was a "case-control study," in which you compare
groups that had the outcome or not, and look at how common the
exposure was in each group. In Denmark, Madsen et al. did the opposite kind of study, called a cohort study: you compare groups that
had the exposure or not, in order to see whether there is any varia-
tion in the outcome. In this specific case, then, you take two groups

of people who either had MMR or didn't and then check later to see if the rate of autism is any different between the two groups.

This study was big—very big—and included all the children born in Denmark between January 1991 and December 1998. In Denmark there is a system of unique personal identification numbers, linked to vaccination registers and information about the diagnosis of autism, which made it possible to chase up almost all the children in the study. This was a pretty impressive achievement, since there were 440,655 children who were vaccinated and 96,648 who were unvaccinated. No difference was found between vaccinated and unvaccinated children in the rates of autism or autistic spectrum disorders and no association between development of autism and age at vaccination.

Anti-MMR campaigners have responded to this work by saying that only a small number of children are harmed by the vaccine, seemingly inconsistent with their claims that MMR is responsible for a massive upswing in diagnoses of autism. In any case, if a vaccine caused an adverse reaction in a very small number of people, that would be no surprise; it would be no different from any other medical intervention (or, arguably, any human activity), and there would be, surely, no story.

As with all studies, there are problems with this huge study. The follow-up of diagnostic records ends one year (December 31, 1999) after the last day of admission to the cohort, so because autism comes on after the age of one year, the children born later in the cohort would be unlikely to have shown up with autism by the end of the follow-up period. But this is flagged up in the study, and you can decide whether you think it undermines its overall findings. I don't think it's much of a problem. That's my verdict, and I think you might agree that it's not a particularly foolish one. It did run from January 1991 after all.

This is the kind of evidence you will find in the Cochrane review, which found, very simply, that "existing evidence on the

safety and effectiveness of MMR vaccine supports current policies of mass immunization aimed at global measles eradication in
order to reduce morbidity and mortality associated with mumps
and rubella."

It also contained multiple criticisms of the evidence it reviewed,
which, bizarrely, have been seized upon by various commentators
to claim that there was some kind of cover-up. The review was
heading toward a conclusion that MMR was risky, they say, if you
read the content, but then, out of nowhere, it produced a reassuring conclusion, doubtless because of hidden political pressure.

The *Daily Mail's* Melanie Phillips, a leading light of the antivaccination movement, was outraged by what she thought she had
found: "It said that no fewer than nine of the most celebrated studies that have been used against [Andrew Wakefield] were unreliable in the way they were constructed." Of course, it did. I'm amazed
it wasn't more. Cochrane reviews are *intended* to criticize papers.

SCIENTIFIC "EVIDENCE" IN THE MEDIA

But the newspapers in 2002 had more than just worried parents.
There was a smattering of science to keep things going: you will
remember computer-generated imagery of viruses and gut walls,
perhaps, and stories about laboratory findings. Why have I not
mentioned those?

For one thing, these important scientific findings were being
reported in newspapers and magazines and at meetings, in fact, anywhere except proper academic journals, where they could be read
and carefully appraised. In May, for example, Wakefield "exclusively revealed" that "more than 95 percent of those who had the
virus in their gut had MMR as their only documented exposure to
measles." He doesn't appear to have revealed this in a peer-reviewed
academic journal, but in a weekend color supplement.

Other people started popping up all over the place, claiming to have made some great finding, but never publishing their research in proper, peer-reviewed academic journals. A pharmacist in Sunderland called Paul Shattock was reported on the *Today* program, and in several national newspapers, to have identified a distinct subgroup of children with autism resulting from MMR. Mr. Shattock is very active on antiimmunization websites, but he still doesn't seem to have got around to publishing this important work years later, even though the Medical Research Council suggested in 2002 that he should "publish his research and come forward to the MRC with positive proposals."

Meanwhile, Dr. Arthur Krigsman, pediatric gastrointestinal consultant working in the New York area, was telling hearings in Washington, D.C., that he had made all kinds of interesting findings in the bowels of autistic children, using endoscopes. This was lavishly reported in the media. Here is *The Daily Telegraph*:

> Scientists in America have reported the first indepen-
> dent corroboration of the research findings of Dr. Andrew
> Wakefield. Dr. Krigsman's discovery is significant because
> it independently supports Dr. Wakefield's conclusion that
> a previously unidentified and devastating combination
> of bowel and brain disease is afflicting young children—
> a claim that the Department of Health has dismissed as
> "bad science."

To the best of my knowledge—and I'm pretty good at searching for this stuff—Krigsman's new scientific research findings that corroborate Andrew Wakefield's have never been published in an academic journal; certainly there is no trace of them on PubMed, the index of nearly all medical academic articles.

In case the reason why this is important has not sunk in, let me explain again. If you visit the premises of the Royal Society in

London, you'll see its motto proudly on display: *Nullius in verba*—
"On the word of no one." What I like to imagine this refers to, in
my geeky way, is the importance of publishing proper scientific
papers if you want people to pay attention to your work. Dr. Arthur
Krigsman has been claiming for years now that he has found evidence linking MMR to autism and bowel disease. Since he hasn't
published his findings, he can claim them until he's blue in the
face, because until we can see exactly what he did, we can't see
what flaws there may be in his methods. Maybe he didn't select
the subjects properly. Maybe he measured the wrong things. If he
doesn't write it up formally, we can never know, because that is
what scientists do: write papers, and pull them apart to see if their
findings are robust.

Krigsman and others' failures to publish in peer-reviewed aca-
demic journals weren't isolated incidents. In fact it's still going on,
years later. In 2006, exactly the same thing was happening again.
US SCIENTISTS BACK AUTISM LINK TO MMR, squealed the *Telegraph*.
SCIENTISTS FEAR MMR LINK TO AUTISM, roared the *Mail*. US STUDY
SUPPORTS CLAIMS OF MMR LINK TO AUTISM, croaked *The Times*
(London) a day later.

What was this frightening new data? These scare stories were
based on a poster presentation, at a conference yet to occur, on research not yet completed, by a man with a track record of announcing research that never subsequently appears in an academic journal.
In fact, astonishingly, four years later, it was Dr. Arthur Krigsman
again. The story this time was different: he had found genetic material (RNA) from vaccine-strain measles virus in some gut samples from children with autism and bowel problems. If true, this
would have fitted with Wakefield's theory, which by 2006 was lying in tatters. We might also mention that Wakefield and Krigsman are doctors together at Thoughtful House, a private autism
clinic in the United States offering eccentric treatments for developmental disorders.

The *Telegraph* went on to explain that Krigsman's most recent unpublished claim was replicating similar work from 1998 by Dr. Andrew Wakefield and from 2002 by Professor John O'Leary. This was, to say the least, a misstatement. There is no work from 1998 by Wakefield that fits the *Telegraph*'s claim—at least not in PubMed that I can find. I suspect the newspaper was confused about the infamous *Lancet* paper on MMR, which by 2004 had already been partially retracted.

There are, however, two papers suggesting that traces of genetic material from the measles virus have been found in children. They have received a mountain of media coverage over half a decade, and yet the media have remained studiously silent on the published evidence suggesting that they were false positives, as we shall now see.

One, in which it is claimed that genetic material from measles vaccine was found in blood cells, is from Kawashima et al. in 2002, also featuring Wakefield as an author. Doubt is cast on this both by attempts to replicate it, showing where the false positives probably appeared, and by the testimony of Nick Chadwick, the Ph.D. student whose work we described above. Even Andrew Wakefield himself no longer relies on this paper.

The other is O'Leary's paper from 2002, also featuring Wakefield as an author, which produced evidence of measles RNA in tissue samples from children. Further experiments, again, have illustrated where the false positives seem to have arisen, and in 2004, when Professor Stephen Bustin was examining the evidence for the legal aid case, he explained how he established to his satisfaction—during a visit to the O'Leary lab—that these were false positives resulting from contamination and inadequate experimental methods. He has shown, first, that there were no controls to check for false positives (contamination is a huge risk when you are looking for minuscule traces of genetic material, so you generally run blank samples to make sure they do come out blank); he found calibration problems with the machines, prob-lems with logbooks, and worse. He expanded on this at enormous length in a U.S. court case on autism and vaccines in 2006. You can read his detailed explanation in full online. To my astonishment, not one journalist in the U.K. has ever bothered to report it.

Both these papers claiming to show a link received blanket media coverage at the time, as did Krigsman's claims.

WHAT THEY DIDN'T TELL YOU

In the May 2006 issue of the *Journal of Medical Virology* there was a very similar study to the one described by Krigsman, only this one had actually been published, by Afzal et al. It looked for mea-sles RNA in children with regressive autism after MMR vaccina-tion, much like the unpublished Krigsman study, and it used tools so powerful they could detect measles RNA down to single-figure copy numbers. It found no evidence of the magic vaccine-strain measles RNA to implicate MMR. Perhaps because of that unfright-ening result, the study was loudly ignored by the press.

Because it has been published in full, I can read it and pick holes in it, and I am more than happy to do so: because science is about critiquing openly published data and methodologies, rather than press-released chimeras, and in the real world all studies have some flaws, to a greater or lesser extent. Often they are practical ones; here, for example, the researchers couldn't get hold of the tis-sue they ideally would have used, because they could not get ethics committee approval for intrusive procedures like lumbar punctures and gut biopsies on children (Wakefield did manage to obtain such samples, but he was, we should remember, found guilty by the U.K. medical regulator (the General Medical Council) of professional misconduct for conducting these invasive tests without ethics com-mittee approval; his sentence has not been handed down as yet).

Surely they could have borrowed some existing samples, from children said to be damaged by vaccines? You'd have thought so. They report in the paper that they tried to ask anti-MMR researchers—if that's not an unfair term—whether they could borrow some of their tissue samples to work on. They were ignored.*

Afzal et al. were not reported in the media, anywhere at all, except by me, in my column.

This is not an isolated case. Another major paper was published in the leading academic journal *Pediatrics* a few months later—to complete media silence—again suggesting very strongly that the earlier results from Kawashima and O'Leary were in error and false positives. D'Souza et al. replicated the earlier experiments very closely, and in some respects more carefully; most important, it traced out the possible routes by which a false positive could have occurred and made some astonishing findings.

False positives are common in PCR, because it works by using enzymes to replicate RNA, so you start with a small amount in your sample, which is then "amplified up," copied over and over again, until you have enough to measure and work with. Beginning with a single molecule of genetic material, PCR can generate one hundred billion similar molecules in an afternoon. Because of this, the PCR process is exquisitely sensitive to contamination—as numerous innocent people languishing in jail could tell you—so you have to be very careful and clean up as you go.

As well as raising concerns about contamination, D'Souza also found that the O'Leary method might have accidentally amplified the wrong bits of RNA.

Let's be clear, this is absolutely not about criticizing individual researchers. Techniques move on, results are sometimes not repli-

*"The groups of investigators that either had access to original autism specimens or investigated them later for measles virus detection were invited to take part in the study but failed to respond. Similarly, it was not possible to obtain clinical specimens of autism cases from these investigators for independent investigations."

cable, and not all double-checking is practical (although Bustin's testimony is that standards in the O'Leary lab were problematic). But what is striking is that the media rabidly picked up on the original frightening data and then completely ignored the new re-assuring data. This study by D'Souza, like Afzal's before it, was unanimously ignored by the media. It was covered, by my count, in: my column, one Reuters piece that was picked up by nobody, and one post on the lead researcher's boyfriend's blog (where he talked about how proud he was of his girlfriend). Nowhere else.*

You could say, very reasonably, that this is all very much par for the course; newspapers report the news, and it's not very inter-esting if a piece of research comes out saying something is safe. But I would argue—perhaps sanctimoniously—that the media have a special responsibility in this case, because they themselves de-manded "more research" and, moreover, because *at the very same time* that they were ignoring properly conducted and fully pub-lished negative findings, they were talking up scary findings from an unpublished study by Krigsman, a man with a track record of making scary claims that remain unpublished.

MMR is not an isolated case in this regard. You might remem-ber the scare stories about mercury fillings from the past two de-cades; they come around every few years, usually accompanied by a personal anecdote in which fatigue, dizziness, and headaches all are vanquished following the removal of the fillings by one vision-ary dentist. Traditionally these stories conclude with a suggestion that the dental establishment may well be covering up the truth about mercury and a demand for more research into its safety.

The first large-scale randomized control trials on the safety of mer-cury fillings were published recently, and if you were waiting to see

---

*In 2008, some journalists deigned—miraculously—to cover a PCR experiment with a negative finding. It was misreported as the definitive refutation of the entire MMR-autism hypothesis. This was a childish overstatement, and that doesn't help anyone, either. I am not hard to please.

these hotly anticipated results, personally demanded by journalists on innumerable newspapers, you'd be out of luck, because they were reported nowhere. Nowhere. This was a study of more than one thousand children, some of whom were given mercury fillings, and some mercury-free fillings, measuring kidney function and neuro-developmental outcomes like memory, coordination, nerve conduction, IQ, and so on over several years. It was a well-conducted study. There were no significant differences between the two groups. That's worth knowing about if you've ever been scared by the media's reports on mercury fillings, and by God, you'd have been scared.

BBC *Panorama* featured a particularly chilling documentary in 1994 titled *The Poison in Your Mouth*. It opened with dramatic footage of men in full protective gear rolling barrels of mercury around. I'm not giving you the definitive last word on mercury here. But I think we can safely assume there is no *Panorama* documentary in the pipeline covering the startling new research data suggesting that mercury fillings may not be harmful after all.

In some respects this is just one more illustration of how unreliable intuition can be in assessing risks like those presented with a vaccine. Not only is it a flawed strategy for this kind of numerical assessment, on outcomes that are too rare for one person to collect meaningful data on them in their personal journey through life; but the information you are fed by the media about the wider population is ludicrously, outrageously, criminally crooked. So at the end of all this, what has the British news media establishment achieved?

OLD DISEASES RETURN

It's hardly surprising that the MMR vaccination rate has fallen from 92 percent in 1996 to 73 percent today. In some parts of London it's down to 60 percent, and figures from 2004 to 2005 showed that in Westminster only 38 percent of children had both jabs by the age of five.*

It is difficult to imagine what could be driving this if not a brilliantly successful and well-coordinated media anti-MMR campaign, which pitched emotion and hysteria against scientific evidence. People listen to journalists: this has been demonstrated repeatedly, and not just with the kinds of stories in this book.

A 2005 study in *The Medical Journal of Australia* looked at mammogram bookings and found that during the peak media coverage of Kylie Minogue's breast cancer, bookings rose by 40 percent. The increase among previously unscreened women in the forty- to sixty-nine-year age-group was 101 percent. These surges were unprecedented. And I'm not cherry-picking: a systematic review from the Cochrane Collaboration found five studies looking at the use of specific health interventions before and after media coverage of specific stories, and each found that favorable publicity was associated with greater use, and unfavorable coverage with lower use.

It's not just the public: medical practice is influenced by the media too, and so are academics. A mischievous paper from *The New England Journal of Medicine* in 1991 showed that if a study was covered by *The New York Times*, it was significantly more likely to be cited by other academic papers. Having come this far, you are probably unpicking this study already. Was coverage in *The New York Times* just a surrogate marker for the importance of the research? History provided the researchers with a control group to compare their results with: for three months, large parts of the paper went on strike, and while the journalists did produce an "edition of record," it was never actually printed. They wrote stories about academic research, using the same criteria to judge importance that

*Not 11.7 percent, as claimed in the *Telegraph* and the *Daily Mail* in February and June 2006.

they always had, but the research they wrote about in articles that never saw the light of day saw no increase in citations.

People read newspapers. Despite everything we think we know, their contents seep in, we believe them to be true, and we act upon them, making it all the more tragic that their contents are so routinely flawed. Am I extrapolating unfairly from the extreme examples in this book? Perhaps not. In 2008 Gary Schwitzer, an ex-journalist who now works on quantitative studies of the media, published an analysis of five hundred health articles covering treatments from mainstream newspapers in the United States. Only 35 percent of stories were rated satisfactory for whether the journalist had "discussed the study methodology and the quality of the evidence" (because in the media, as we have seen repeatedly in this book, science is about absolute truth statements from arbitrary authority figures in white coats, rather than clear descriptions of studies and the reasons why people draw conclusions from them). Only 28 percent adequately covered benefits, and only 33 percent adequately covered harms. Articles routinely failed to give any useful quantitative information in absolute terms, preferring unhelpful eye catchers like "50 percent higher" instead.

In fact, there have been systematic quantitative surveys of the accuracy of health coverage in Canada, Australia, and the United States—I'm trying to get one off the ground in the U.K.—and the results have been universally unimpressive. It seems to me that the state of health coverage in the U.K. could well be a serious public health issue.

Meanwhile, the incidence of two of the three diseases covered by MMR is now increasing very impressively. We have the highest number of measles cases in England and Wales since current surveillance methods began in 1995, with cases occurring mostly in children who had not been adequately vaccinated: in 2007, 971 confirmed cases (mostly associated with prolonged outbreaks in traveling and religious communities, where vaccine uptake has been historically low), were reported, after 740 cases in 2006 (and the first death since 1992). Seventy-three percent of cases were in the southeast, and most of those were in London.

Mumps began rising again in 1999, after many years of cases in only double figures; by 2005 the United Kingdom had a mumps epidemic, with around five thousand notifications in January alone.

A lot of people who campaign against vaccines like to pretend that they don't do much good and that the diseases they protect against were never very serious anyway. I don't want to force anyone to have his or her child vaccinated, but equally I don't think anyone is helped by misleading information. By contrast with the unlikely event of autism's being associated with MMR, the risks from measles, though small, are real and quantifiable. The Peckham Report on immunization policy, published shortly after the introduction of the MMR vaccine, surveyed the recent experience of measles in Western countries and estimated that for every 1,000 cases notified, there would be 0.2 deaths, 10 hospital admissions, 10 neurological complications, and 40 respiratory complications. These estimates have been borne out in recent minor epidemics in the Netherlands (1999: 2,300 cases in a community philosophically opposed to vaccination, 3 deaths), Ireland (2000: 1,200 cases, 3 deaths) and Italy (2002: 3 deaths). It's worth noting that plenty of these deaths were in previously healthy children, in developed countries, with good health care systems.

Though mumps is rarely fatal, it's an unpleasant disease with unpleasant complications (including meningitis, pancreatitis, and sterility). Congenital rubella syndrome has become increasingly rare since the introduction of MMR but causes profound disabilities, including deafness, autism, blindness, and mental handicap, resulting from damage to the fetus during early pregnancy.

The other thing you will hear a lot is that vaccines don't make much difference anyway, because all the advances in health and life expectancy have been due to improvements in public health for a wide range of other reasons. As someone with a particular interest in epidemiology and public health, I find this suggestion flattering, and there is absolutely no doubt that deaths from measles began to fall over the whole of the past century for all kinds of reasons, many of them social and political as well as medical: better nutrition, better access to good medical care, antibiotics, less crowded living conditions, improved sanitation, and so on.

Life expectancy in general has soared over the past century, and it's easy to forget just how phenomenal this change has been. In 1901, males born in the U.K. could expect to live to forty-five, and females to forty-nine. By 2004, life expectancy at birth had risen to seventy-seven for men and eighty-one for women (although of course, much of the change is due to reductions in infant mortality).

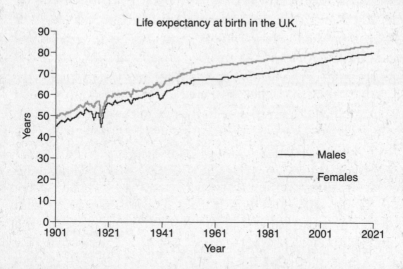

So we are living longer, and vaccines are clearly not the only reason why. No single change is the reason why. Measles incidence dropped hugely over the preceding century, but you would have to work fairly hard to persuade yourself that vaccines had no impact on that. Here, for example, is a graph showing the reported incidence of measles from 1950 to 2000 in the United States.

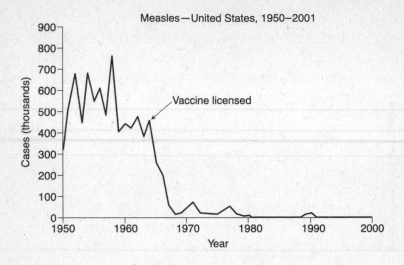

Measles—United States, 1950–2001

For those who think that single vaccines for the components of MMR are a good idea, you'll notice that these have been around since the 1970s, but that a concerted program of vaccination, and the concerted program of giving all three vaccinations in one go as MMR, are fairly clearly associated in time with a further (and actually rather definitive) drop in the rate of measles cases.

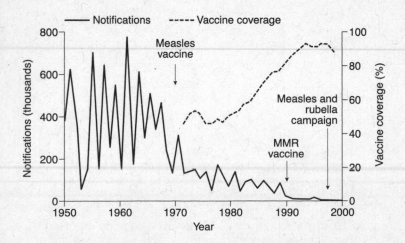

The same is true for mumps.

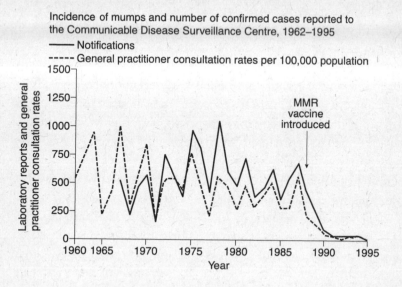

Incidence of mumps and number of confirmed cases reported to
the Communicable Disease Surveillance Centre, 1962–1995
—— Notifications
----- General practitioner consultation rates per 100,000 population

While we're thinking about mumps, let's not forget our epidemic in 2005, a resurgence of a disease that many young doctors would struggle even to recognize. Here is a graph of mumps cases from the *BMJ* article that analyzed the outbreak:

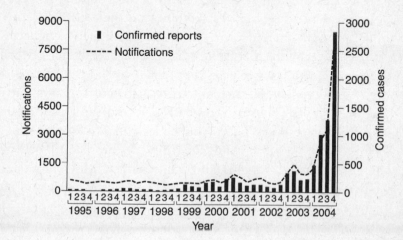

Almost all confirmed cases during this outbreak were in people aged fifteen to twenty-four, and only 3.3 percent had received the full two doses of MMR vaccine. Why did it affect these people? Because of a global vaccine shortage in the early 1990s.

Mumps is not a harmless disease. I've no desire to scare anyone—and as I said, your beliefs and decisions about vaccines are your business; I'm interested only in how you came to be so incredibly misled—but before the introduction of MMR, mumps was the commonest cause of viral meningitis and one of the leading causes of hearing loss in children. Lumbar puncture studies show that around half of all mumps infections involve the central nervous system. Mumps orchitis is common, exquisitely painful, and occurs in 20 percent of adult men with mumps; around half will experience testicular atrophy, normally in only one testicle, but 15 to 30 percent of patients with mumps orchitis will have it in both testicles, and of these, 13 percent will have reduced fertility.

I'm not just spelling this out for the benefit of the lay reader; by the time of the outbreak in 2005, young doctors needed to be reminded of the symptoms and signs of mumps, because it had been such an uncommon disease during their training and clinical experience. People had forgotten what these diseases looked like, and in that regard vaccines are a victim of their own success, as we saw in our earlier quote from *Scientific American* in 1888, five generations ago (see page 214).

Whenever we take a child to be vaccinated, we're aware that we are striking a balance between benefit and harm, as with any medical intervention. I don't think vaccination is all that important: even if mumps orchitis, infertility, deafness, death, and the rest are no fun, the sky wouldn't fall in without MMR. But taken on their own, lots of other individual risk factors aren't very important either, and that's no reason to abandon all hope of trying to do something simple, sensible, and proportionate about them, gradually increasing the health of the nation, along with all the other stuff you can do to the same end.

It's also a question of consistency. At the risk of initiating mass panic, I feel duty bound to point out that if MMR still scares you, then so should everything in medicine and indeed many of the everyday lifestyle risk exposures you encounter, because there are a huge number of things that are far less well researched, with a far lower level of certainty about their safety. The question would still remain of why you were so focused on MMR. If you wanted to do something constructive about this problem, instead of running a single-issue campaign about MMR, you might, per-haps, use your energies more usefully. You could start a campaign for constant automated vigilance of the entirety of the Food and Drug Administration data set for any adverse outcomes associ-ated with any intervention, for example, and I'd be tempted to join you on the barricades.

But in many respects this isn't about risk management or vigi-lance; it's about culture, human stories, and everyday human harms. Just as autism is a peculiarly fascinating condition to journalists, and indeed to all of us, vaccination is similarly inviting as a focus for our concerns. It's a universal program, in conflict with modern ideas of "individualized care"; it's bound up with government; it in-volves needles going into children; and it offers the opportunity to blame someone, or something, for a dreadful tragedy.

Just as the causes of these scares have been more emotional than anything else, so too has much of the harm. Parents of chil-dren with autism have been racked with guilt, doubt, and endless self-recrimination over the thought that they themselves are re-sponsible for inflicting harm upon their own children. This distress has been demonstrated in countless studies: but so close to the end, I don't want to introduce any more research papers.

There is one quote that I find, although she would perhaps complain about my using it, both moving and upsetting. It's from

Karen Prosser, who featured with her autistic son Ryan in the Andrew Wakefield video news release from the Royal Free Hospital in 1998. "Any mother who has a child wants it to be normal," she says. "To then find out your child might be genetically autistic is tragic. To find out that it was caused by a vaccine, that you agreed to have done . . . is just devastating."

AND ANOTHER THING

I could go on. As I write this, the media are still pushing a celebrity-endorsed "miracle cure" (and I quote) for dyslexia, invented by a millionaire paint entrepreneur, despite the abysmal evidence to support it, and despite customers' being at risk of simply losing their money anyway, because the company seems to be going into administration; more "hidden data" scandals are exposed from the vaults of big pharma every month; quacks and cranks continue to parade themselves on television, quoting fantastical studies to universal approbation; and there will always be new scares, because they sell so very well, and they make journalists feel alive.

To anyone who feels their ideas have been challenged by this book or who has been made angry by it—to the people who feature in it, I suppose—I would say this: you win. You really do. I would hope there might be room for you to reconsider, to change your stance in the light of what might be new information (as I will happily do, if there is ever an opportunity to update this book). But you will not need to because, as we both know, you collectively have almost full-spectrum dominance. Your ideas—bogus though they may be—have immense superficial plausibility, they can be expressed rapidly, they are endlessly repeated, and they are believed by enough people for you to make very comfortable livings and to have enormous cultural influence. You win.

It's not the spectacular individual stories that are the problem so much as the constant daily grind of stupid little ones. This will not end, and so I will now abuse my position by telling you, very briefly, exactly what I think is wrong and some of what can be done to fix it.

The process of obtaining and interpreting evidence isn't taught in schools, nor are the basics of evidence-based medicine and epidemiology, yet these are obviously the scientific issues that are most on people's minds. Science coverage now tends to come from the world of medicine, and the stories are of what will kill you, or save you. Perhaps it is narcissism, or fear, but the science of health is important to people, and at the very time when we need it the most, our ability to think around the issue is being energetically distorted by the media, corporate lobbies, and, frankly, cranks.

Without anybody's noticing, bullshit has become an extremely important public health issue, and for reasons that go far beyond the obvious hysteria around immediate harms, the odd measles tragedy or a homeopath's unnecessary malaria case. Doctors today are keen—as it said in our medical school notes—to work "collaboratively with the patient toward an optimum health outcome." They discuss evidence with their patients, so that they can make their own decisions about treatments.

In my job as a doctor I meet patients from every conceivable walk of life, in huge numbers, discussing some of the most important issues in their lives. This has consistently taught me one thing: people aren't stupid. Anybody can understand anything, as long as it is clearly explained, but, more than that, if they are sufficiently interested. What determines an audience's understanding is not so much scientific knowledge as motivation: patients who are ill, with an important decision to make about treatment, can be very motivated indeed.

But journalists and miracle cure merchants sabotage this process of shared decision making, diligently, brick by brick, making lengthy and bogus criticisms of the process of systematic review (because they don't like the findings of just one), extrapolating from lab dish data, misrepresenting the sense and value of trials, carefully and collectively undermining people's understanding of the very notion of what it means for there to be evidence for an activity. In this regard they are, to my mind, guilty of an unforgivable crime.

You'll notice, I hope, that I'm more interested in the cultural impact of nonsense—the medicalization of everyday life, the undermining of sense—and in general I blame systems more than particular people. While I do go through the backgrounds of some individuals, this is largely to illustrate the extent to which they have been misrepresented by the media, who are so desperate to present their favored authority figures as somehow mainstream. I am not surprised that there are individual entrepreneurs, but I am unimpressed that the media carry their assertions as true. I am not surprised that there are people with odd ideas about medicine or that they sell those ideas. But I am spectacularly, supremely, incandescently unimpressed when a university starts to offer B.Sc. science courses in them. I do not blame individual journalists (for the most part), but I do blame whole systems of editors and the people who buy newspapers with values they profess to despise.

Similarly, while I could reel out a few stories of alternative therapists' customers who've died unnecessarily, it seems to me that people who choose to see alternative therapists (except for nutrition therapists, who have worked *very* hard to confuse the public and to brand themselves as conventional evidence-based practitioners) make that choice with their eyes open or at least only half closed. To me this is not a situation of businesspeople exploiting the vulnerable, but rather, as I seem to keep saying, a bit more complicated than that. We love this stuff, and we love it for some fascinating reasons, which we could ideally spend a lot more time thinking and talking about.

Economists and doctors talk about "opportunity costs," the things you could have done but didn't, because you were distracted by doing something less useful. To my mind, the greatest harm posed by the avalanche of nonsense we have seen in this book is best conceived of as the "opportunity cost of bullshit."

We have somehow become collectively obsessed with these absurd, thinly evidenced individual tinkerings in diet, distracting us from simple, healthy eating advice but, more than that, as we saw, distracting us from the other important lifestyle risk factors for ill health that cannot be sold or commodified.

Doctors, similarly, have been captivated by the commercial success of alternative therapists. They could learn from the best of the research into the placebo effect and the meaning response in healing, and apply that to everyday clinical practice, augmenting treatments that are in themselves also effective; instead there is a fashion among huge numbers of them to indulge childish fantasies about magic pills, massages, or needles. That is not forward-looking or inclusive, and it does nothing about the untherapeutic nature of rushed consultations in decaying buildings. It also requires, frequently, that you lie to patients. "The true cost of something," as *The Economist* says, "is what you give up to get it."

On a larger scale, many people are angry about the evils of the pharmaceutical industry and nervous about the role of profit in health care; but these are formless and uncalibrated intuitions, so the valuable political energy that comes from this outrage is funneled—wasted—through infantile issues like the miraculous properties of vitamin pills or the evils of MMR. Just because big pharma can behave badly, that does not mean that sugar pills work better than placebo, nor does it mean that MMR causes autism. Whatever the wealthy pill peddlers try to tell you, with their brand-building conspiracy theories, big pharma isn't *afraid* of the food supplement pill industry; it *is* the food supplement pill industry. Similarly, big pharma isn't frightened for its profits because popular opinion turned against MMR: if they have any sense, these companies are relieved that the public is obsessed with MMR and thus distracted from the other far more complex and real issues connected with the pharmaceutical business and its inadequate regulation.

To engage meaningfully in a political process of managing the evils of big pharma, we need to understand a little about the business of evidence; only then can we understand why transparency is so important in pharmaceutical research, for example, or the details of how it can be made to work or concoct new and imaginative solutions.

But the greatest opportunity cost comes, of course, in the media, who have failed science so spectacularly, getting stuff wrong and dumbing down. No amount of training will ever improve the wildly inaccurate stories, because newspapers already have specialist health and science correspondents who understand science. Editors will always—cynically—sideline those people and give stupid stories to generalists, for the simple reason that they want stupid stories. Science is beyond their intellectual horizon, so they assume you can just make it up anyway. In an era when mainstream media are in fear for their lives, their claims to act as effective gatekeepers to information are somewhat undermined by the content of pretty much every column or blog entry I've ever written.

To academics, and scientists of all shades, I would say this: you cannot ever possibly prevent newspapers from printing nonsense, but you can add your own sense into the mix. E-mail the features desk, ring the health desk (you can find the switchboard number on the letters page of any newspaper), and offer them a piece on something interesting from your field. They'll turn you down. Try again. You can also toe the line by not writing stupid press releases (there are extensive guidelines for communicating with the media online), by being clear about what's speculation in your discussions, by presenting risk data as "natural frequencies," and so on. If you feel your work—or even your field—has been misrepresented, then complain. Write to the editor, the journalist, the letters page, the readers' editor, start a blog, put out a press release explaining why the story was stupid; get your press office to harass the paper or TV station, use your title (it's embarrassing how easy they are to impress), and offer to write them something yourself.

The greatest problem of all is dumbing down. Everything in the media is robbed of any scientific meat, in a desperate bid to seduce an imaginary mass that aren't interested. And why should they be? Meanwhile, the nerds, the people who studied biochemistry but now work in middle management, are neglected, unstimulated, abandoned. There are intelligent people out there who want to be pushed, to keep their knowledge and passion for science alive, and neglecting them comes at a serious cost to society. Institutions have failed in this regard. The indulgent and well-financed "public engagement with science" community has been worse than useless, because it too is obsessed with taking the message to everyone, rarely offering stimulating content to the people who are already interested.

Now you don't need these people. Start a blog. Not everyone will care, but some will, and they will find your work. Unmediated access to niche expertise is the future, and you know, science isn't hard—academics around the world explain hugely complicated ideas to ignorant eighteen-year-olds every September—it just requires motivation. I give you the CERN podcast, the Science in the City mp3 lecture series, blogs from profs, open-access academic journal articles from PLOS, online video archives of popular lectures, the free editions of the Royal Statistical Society's magazine *Significance*, and many more, all out there, waiting for you to join them. There's no money in it, but you knew that when you started on this path. You will do it because you know that knowledge is beautiful, and because if only a hundred people share your passion, that is enough.

NOTES

1. MATTER

4   "The first day": *New York Daily News* (September 30, 2007).

8   "The candles work by": www.bbc.co.uk/wales/southeast/sites/mind/pages/hopi.shtml.

8   a paper published: Seely, D. R.; Quigley, S. M., and Langman, A. W. Ear candles—efficacy and safety. *Laryngoscope* 106, no. 10 (October 1996): 1226–29.

9   a published study: Ibid.

13   "These cleansing and": Green, E. C., and Honwana, A. Indigenous healing of war-affected children in Africa. IK Notes No. 10. Knowledge and Learning Center Africa Region, World Bank Washington (1999); available: www.africaaction.org/docs99/viol9907.htm.

4. HOMEOPATHY

40   In one study: Marshall, T. Reducing unnecessary consultation—a case of NNNT? *Bandolier* 44, no. 4 (1997): 1–3.

42   But the point of the study: MacManus, M. P.; Matthews, J. P.; Wada, M.; Wirth, A.; Worotniuk, V.; Ball, D. L. Unexpected long-term survival after low-dose palliative radiotherapy for non-small cell lung cancer. *Cancer* 106, no. 5 (March 1, 2006): 1110–16.

48   a very theatrical trial: Majeed, A. W., et al. Randomised, prospective, single-blind comparison of laparoscopic versus small-incision cholecystectomy. *Lancet* 347, no. 9007 (April 13, 1996): 989–94.

48   a review of blinding: Schulz, K. F.; Chalmers, I.; Hayes, R. J.; Altman, D. G. Empirical evidence of bias: dimensions of methodological quality associated with estimates of treatment effects in controlled trials. *JAMA* 273 (1995): 408–12.

48   a review of trials: Ernst, E., and White, A. R. Acupuncture for back pain: a meta-analysis of randomized controlled trials. *Arch Int Med* 158 (1998): 2235–41.

49   Diagram: Schulz et al. Empirical evidence of bias, 408–12.

49   a standard review: Ernst, E., and Pittler, M. H. Efficacy of homeopathic arnica: a systematic review of placebo-controlled clinical trials. *Arch Surg.* 133, no. 11 (November 1998): 1187–90.

50   "Let us take out": van Helmont, J. B. *Oriatrike, or Physick Refined: The Common Errors Therein Refuted and the Whole Are Reformed and Rectified* (London: Lodowick-Loyd, 1662), p. 526; available at www.jameslindlibrary .org.

52   two landmark studies: Khan, K. S.; Daya, S.; Jadad, A. R. The importance of quality of primary studies in producing unbiased systematic reviews. *Arch Intern Med* 156 (1996): 661–66; Moher, D.; Pham, B.; Jones, A., et al. Does quality of reports of randomized trials affect estimates of intervention efficacy reported in meta-analyses? *Lancet* 352 (1998): 609–13.

55   Diagram: Ernst, E., and Pittler, M. H. Re-analysis of previous meta-analysis of clinical trials of homeopathy. J Clin Epi 53, no. 11 (2000): 1188.

58   A landmark meta-analysis: Shang, A.; Huwiler-Müntener, K.; Nartey, L.; Jüni, P.; Dörig, S.; Sterne, J. A.; Pewsner, D.; Egger, M. Are the clinical effects of homoeopathy placebo effects? Comparative study of placebo-controlled trials of homoeopathy and allopathy. *Lancet* 366, no. 9487 (August 27, September 2, 2005): 726–32.

60   One study actually thought: Tallon, D.; Chard, J.; Dieppe, P. Relation between agendas of the research community and the research consumer. *Lancet* 355 (2000): 2037–40.

62   "They might flick": BBC Radio 4 Case Notes, July 19, 2005.

5. THE PLACEBO EFFECT

65   "Shall [the placebo]": The placebo in medicine. *Medical Press* (June 18, 1890): 642.

66   Henry Beecher: Beecher, H. K. The powerful placebo. *JAMA* 159, no. 17 (December 24, 1955): 1602–06.

66   Peter Parker: Skrabanek, P., and McCormick, J. *Follies and Fallacies in Medicine* (Glasgow: Tarragon Press, 1989).

68   Daniel Moerman: Moerman, D. E. General medical effectiveness and human biology: placebo effects in the treatment of ulcer disease. *Med Anth Quarterly* 14, no. 4 (August 1983): 3–16.

68 in a different data set: de Craen, A. J.; Moerman, D. E.; Heisterkamp, S. H.; Tytgat, G. N.; Tijssen, J. G.; Kleijnen, J. Placebo effect in the treatment of duodenal ulcer. *Br J Clin Pharmacol* 48, no. 6 (December 1999): 853–60.

69 Blackwell (1972): Blackwell, B.; Bloomfield, S. S.; Buncher, C. R. Demonstration to medical students of placebo responses and non-drug factors. *Lancet* 1, no. 7763 (June 10, 1972): 1279–82.

69 Another study: Schapira, K.; McClelland, H. A.; Griffiths, N. R.; Newell, D. J. Study on the effects of tablet color in the treatment of anxiety states. *BMJ* 1, no. 5707 (May 23, 1970): 446–49.

69 a survey of the color: de Craen, A. J.; Roos, P. J.; Leonard de Vries, A.; Kleijnen, J. Effect of color of drugs: systematic review of perceived effect of drugs and of their effectiveness. *BMJ* 313, no. 7072 (December 21–28, 1996): 1624–26.

70 In 1970: Hussain, M. Z., and Ahad, A. Tablet color in anxiety states. *BMJ* 3, no. 5720 (August 22, 1970): 466.

70 Route of administration: Grenfell, R. F.; Briggs, A. H.; Holland, W. C. Double blind study of the treatment of hypertension. *JAMA* 176 (1961): 124–28; De Craen, A.J.M.; Tijssen, J.G.P.; de Gans, J.; Kleijnen, J. Placebo effect in the acute treatment of migraine: subcutaneous placebos are better than oral placebos. *J Neur* 247 (2000): 183–88; Gracely, R. H.; Dubner, R.; McGrath, P. A. Narcotic analgesia: fentanyl reduces the intensity but not the unpleasantness of painful tooth pulp sensations. *Science* 203, no. 4386, (March 23, 1979): 1261–63.

70 the bizarre story of packaging: Kaptchuk, T. J.; Stason, W. B.; Davis, R. B.; Legedza, A. R.; Schnyer, R. N.; Kerr, C. E.; Stone, D. A.; Nam, B. H.; Kirsch, I.; Goldman, R. H. Sham device v inert pill: randomized controlled trial of two placebo treatments. *BMJ* 332, no. 7538 (February 18, 2006): 391–97.

70 Branthwaite and Cooper: Branthwaite, A., and Cooper, P. Analgesic effects of branding in treatment of headaches. *BMJ* (Clin Res ed.) 282 (1981): 1576–78.

71 a recent study: Waber et al. Commercial features of placebo and therapeutic efficacy. *JAMA* 299 (2008): 1016–17.

71 a paper currently in press: Ginoa, F. Do we listen to advice just because we paid for it? The impact of advice cost on its use. Organizational Behavior and Human Decision Processes, Vol. 107, Issue 2, November 2008, 234 45 (published online April 25, 2008), dx.doi.org/10.1016/j.obhdp.2008 .03.001.

71 Montgomery and Kirsch: Montgomery, G. H., and Kirsch, I. Mechanisms of placebo pain reduction: an empirical investigation. *Psych Science* 7 (1996): 174–76.

72 a placebo-controlled trial: Cobb, L. A.; Thomas, G. I.; Dillard, D. H.; Merendino, K. A.; Bruce, R. A. An evaluation of internal-mammary-artery

ligation by a double-blind technic. *N Eng J Med* 260, no. 22 (May 28, 1959): 1115–18.

73 A Swedish study: Linde, C.; Gadler, F.; Kappenberger, L.; Rydén, L. Placebo effect of pacemaker implantation in obstructive hypertrophic cardiomyopathy. PIC Study Group. *Am J Cardiol* 83, no. 6 (March 15, 1999): 903–37.

73 "Electrical machines have": Johnson, A. G. Surgery as a placebo. *Lancet* 344, no. 8930 (October 22, 1994): 1140–42.

73 an elegant study: Crum, A. J., and Langer, E. J. Mind-set matters: exercise and the placebo effect. *Psych Science* 18, no. 2 (February 2007): 165–71.

74 Gryll and Katahn: Gryll, S. L., and Katahn, M. Situational factors contributing to the placebos effect. *Psychopharmacology* (Berlin) 57 (1978): 253–61.

75 Gracely (1985): Gracely, R. H.; Dubner, R.; Deeter, W. R.; Wolskee, P. J. Clinicians' expectations influence placebo analgesia. *Lancet* 1, no. 8419 (January 5, 1985): 43.

76 In 1987, Thomas: Thomas, K. B. General practice consultations: is there any point in being positive? *BMJ* (Clin Res ed.) 294, no. 6581 (May 9, 1987): 1200–02.

77 Raymond Tallis: Tallis, R. *Hippocratic Oaths: Medicine and Its Discontents.* (New York: Atlantic, 2004).

78 Quesalid: Lévi-Strauss, C. The sorcerer and his magic. In *Structural Anthropology,* trans Jacobson, C., and Schoef, B. G. (New York: Basic Books, 1963).

79 a classic study from 1965: Park, L. C., and Covi, L. Nonblind placebo trial: an exploration of neurotic patients' responses to placebo when its inert content is disclosed. *Arch Gen Psych* 12 (April 1965): 36–45.

80 Dr. Stewart Wolf: Wolf, S. Effects of suggestion and conditioning on the action of chemical agents in human subjects; the pharmacology of placebos. *J Clin Invest,* 29, no. 1 (January 1950): 100–09.

80 It's been shown: de la Fuente-Fernández, R.; Ruth, T. J.; Sossi, V.; Schulzer, M.; Calne, D. B.; Stoessl, A. J. Expectation and dopamine release: mechanism of the placebo effect in Parkinson's disease. *Science* 293, no. 5532 (August 10, 2001): 1164–66.

80 Zubieta (2005): Zubieta, J. K.; Bueller, J. A.; Jackson, L. R.; Scott, D. J.; Xu, Y.; Koeppe, R. A.; Nichols, T. E.; Stohler, C. S. Placebo effects mediated by endogenous opioid activity on mu-opioid receptors. *J Neur* 25, no. 34 (August 24, 2005): 7754–62.

81 Researchers have measured: Ader, R., and Cohen, N. Behaviorally conditioned immunosuppression. *Psychosom Med* 37, no. 4 (July–August 1975): 333–40.

81 researchers gave healthy: Goebel, M. U.; Trebst, A. E.; Steiner, J.; Xie, Y. F.; Exton, M. S.; Frede, S.; Canbay, A. E.; Michel, M. C.; Heemann, U.;

Schedlowski, M. Behavioral conditioning of immunosuppression is possible in humans. *FASEB J* 16, no. 14 (December 2002): 1869–73.

81 Researchers have even: Buske-Kirschbaum; A.; Kirschbaum, C.; Stierle, H.; Lehnert, H.; Hellhammer, D. Conditioned increase of natural killer cell activity (NKCA) in humans. *Psychosom Med* 54, no. 2 (March–April 1992): 123–32.

81 From survey data: Goodwin, J. S.; Goodwin, J. M.; Vogel, A. V. Knowledge and use of placebos by house officers and nurses. *Ann Intern Med* 91, no. 1 (July 1979): 106–10.

81 You are a placebo responder: Moerman, D. E. *Meaning, Medicine and the "Placebo Effect"* (New York: Cambridge University Press, 2002), 34, summarizing secondary references to five further studies.

82 one of the most impressive: Moerman, D. E., and Harrington, A. Making space for the placebo effect in pain medicine. *Sem in Pain Med* 3, 1 spec issue (March 2005): 2–6.

82 Another study from 2002: Walsh, B. T.; Seidman, S. N.; Sysko, R.; Gould, M. Placebo response in studies of major depression: variable, substantial, and growing. *JAMA*, 287, no. 14 (10 April 2002): 1840–47.

85 one study found: Ernst, E., and Schmidt, K. Aspects of MMR. *BMJ* 325 (2002): 597.

6. THE NONSENSE DU JOUR

89 "It is impossible": Frankfurt, H. G. *On Bullshit* (Princeton, N.J.: Princeton University Press, 2005); press.princeton.edu/video/frankfurt.

105 over fifty billion: www.nutraingredientsusa.com/news/ng.asp?n=85087.

105 One was in Finland: Alpha-Tocopherol Beta-Carotene Cancer Prevention Study Group. The effect of vitamin E and beta-carotene on the incidence of lung and other cancers in male smokers. *NEJM* 330 (1994): 1029–35.

106 Two groups of people: Thornquist, M. D.; Omenn, G. S.; Goodman, G. E.; Grizzle, J. E.; Rosenstock, L.; Barnhart, S.; Anderson, G. L.; Hammar, S.; Balmes, J.; Cherniack, M. Statistical design and monitoring of the Carotene and Retinol Efficacy Trial (CARET). *Control Clin Trials* 14 (1993): 308–24; Omenn, G. S.; Goodman, G. E.; Thornquist, M. D.; Balmes, J.; Cullen, M. R.; Glass, A., et al. Effects of a combination of beta-carotene and vitamin A on lung cancer and cardiovascular disease. *N Engl J Med* 334 (1996): 1150–55; jnci.oxfordjournals.org/cgi/ijlink?linkType=ABST& journalCode=nejm&resid=334/18/1150].

107 The most up-to-date Cochrane: Vivekananthan, D. P., et al. Use of antioxidant vitamins for the prevention of cardiovascular disease: meta-analysis of randomized trials. *Lancet* 361 (2003): 2017–23; www.thelancet.com/ journals/lancet/article/PIIS0140673603136379/abstract.

107 The Cochrane review: Caraballoso, M.; Sacristan, M.; Serra, C.; Bonfill, X. Drugs for preventing lung cancer in healthy people. *Cochrane Database of Systematic Reviews* 2 (2003).

107 a Cochrane review: Bjelakovic, G.; Nikolova, D.; Gluud, L. L.; Simonetti, R. G.; Gluud, C. Antioxidant supplements for prevention of mortality in healthy participants and patients with various diseases. *Cochrane Database of Systematic Reviews* 2 (2008).

108 Dr. Benjamin Spock: Chalmers, I. Invalid health information is potentially lethal. *BMJ* 322, no. 7292 (2001): 998.

109 price-fixing cartel: Kluwer, John M. Connor, *Global Pricefixing: Our Customers Are the Enemy*, (New York: Springer, 2001); available online, books.google.co.uk/books?id=7M8n4UN23WsC.

109 "Doubt is our product": Michaels David, ed., *Doubt Is Their Product: How Industry's Assault on Science Threatens Your Health* (New York: Oxford University Press, 2008).

7. NUTRITIONISTS

114 Dudley J. LeBlanc: Anderson, Ann, *Snake Oil, Hustlers and Hambones: The American Medicine Show* (Jefferson, N.C.: McFarland, 2005).

121 "During the war": Commencement Speech from Caltech 1974, also in Richard Feynman, Richard, *Surely You're Joking, Mr. Feynman!: Adventures of a Curious Character*. (New York: W. W. Norton, 1985).

121 Clayton College of Natural Health website: www.ccnh.edu/about/programs/tuition.aspx.

8. THE DOCTOR WILL SUE YOU NOW

134 One study estimates: Nattrass, N. Estimating the lost benefits of antiretroviral drug use in South Africa. *African Affairs* 107, no. 427 (2008): 157–76.

135 Another study: Chigwedere, P.; Seage, G. R.; Gruskin, S.; Lee, T. H.; Essex, M. Estimating the lost benefits of antiretroviral drug use in South Africa. *Journal of Acquired Immune Deficiency Syndromes* 49, no. 4 (December 1, 2008): 410–15.

136 "gave a presentation": www.villagevoice.com/2000-0704/news/debating-the-obvious.

9. IS MAINSTREAM MEDICINE EVIL?

148 From the state of current knowledge: clinicalevidence.bmj.com/ceweb/about/knowledge.jsp.

148 These real-world studies: The classic general medicine reference for this is Ellis, J.; Mulligan, I.; Rowe, J.; Sackett, D. L. Inpatient general medicine is evidence based. A-Team, Nuffield Department of Clinical Medicine. *Lancet* 346, no. 8972 (August 12, 1995): 407–10. There have been numerous copycat studies in various specialties, and rather than list them here, an excellent review of them is maintained at www.shef.ac.uk/scharr/ir/percent.html.

149 all those studies: Mayor, S. Audit identifies the most read BMJ research papers. *BMJ* 334 (2007): 554–55; Hippisley-Cox, J., and Coupland, C. Risk of myocardial infarction in patients taking cyclo-oxygenase-2 inhibitors or conventional nonsteroidal anti-inflammatory drugs: population based nested case-control analysis. *BMJ* 330 (2005): 1366; Gunnell, J.; Saperia, J.; Ashby, D. Selective serotonin reuptake inhibitors (SSRIs) and suicide in adults: meta-analysis of drug company data from placebo controlled, randomized controlled trials submitted to the MHRA's safety review. *BMJ* 330 (2005): 385; Fergusson, D., et al. Association between suicide attempts and selective serotonin reuptake inhibitors: systematic review of randomized controlled trials. *BMJ* 330 (2005): 396.

149 annual spend on promotion: Hollon, M. F. Direct-to-consumer advertising: a haphazard approach to health promotion. *JAMA* 293, no. 16 (April 27, 2005): 2030–33. doi:10.1001/jama.293.16.2030. jama.ama-assn.org.

153 whole areas can be orphaned: Iribarne, A. Orphan diseases and adoptive initiatives. *JAMA* 290 (2003): 116; Francisco, A. Drug development for neglected diseases. *Lancet* 360 (2002): 1102.

155 If you follow the references: Safer, D. J. Design and reporting modifications in industry-sponsored comparative psychopharmacology trials. *J Nerv Ment Dis* 190 (2002): 583–92.

156 various studies have shown: Modell et al. (1997); Montejo-Gonzalez et al. (1997); Zajecka et al. (1999); Preskorn (1997): in Safer, ibid.

158 If the difference: Pocock, S. J. When (not) to stop a clinical trial for benefit. *JAMA*, 294 (2005): 2228–30.

159 a systematic review: Lexchin, J.; Bero, L. A.; Djulbegovic, B.; Clark, O. Pharmaceutical industry sponsorship and research outcome and quality. *BMJ* 326 (2003): 1167–70.

159 One review of bias: Rochon, P. A.; Gurwitz, J. H.; Simms, R. W.; Fortin, P. R.; Felson, D. T.; Minaker, K. L.; Chalmers, T. C. A study of manufacturer-supported trials of nonsteroidal anti-inflammatory drugs in the treatment of arthritis. *Arch Intern Med.* 154, no. 2 (January 24, 1994): 157–63.

160 when the methodological flaws: Lexchin, J.; Bero, L. A.; Djulbegovic, B.; Clark, O. Pharmaceutical industry sponsorship and research outcome and quality: systematic review. *BMJ* 326, no. 7400 (May 31, 2003): 1167–70.

162 In 1995, only: Schmidt, K.; Pittler, M. H.; Ernst, E. Bias in alternative medicine is still rife but is diminishing. *BMJ* 323, no. 7320 (November 3, 2001): 1071.

162 A review in 1998: Vickers, A.; Goyal, N.; Harland, R.; Rees, R. Do certain countries produce only positive results? A systematic review of controlled trials. *Control Clin Trials* 19, no. 2 (April 1998): 159–66.

163 a paper has even found: Dubben, H., and Beck-Bornholdt, H. Systematic review of publication bias in studies on publication bias. *BMJ* 331 (2005): 433–34.

163 published a paper: Turner, E. H.; Matthews, A. M.; Linardatos, E.; Tell, R. A.; Rosenthal, R. Selective publication of antidepressant trials and its influence on apparent efficacy. *N Eng J Med* 358, no. 3 (January 17, 2008): 252–60.

164 A classic piece of detective work: Tramer, M. R.; Reynolds, D.J.M.; Moore, R. A.; McQuay, H. J. Impact of covert duplicate publication on meta-analysis: a case study. *BMJ* 315 (1997): 635–40.

165 "When we carried out": Cowley, A. J. et al. *Int Journ Card* 40 (1993): 161–66.

166 the three highest-ranking papers: Audit identifies the most read BMJ research papers. *BMJ* 334 (March 17, 2007): 554–55.

167 "It is a shame": Scolnick, E. M. E-mail communication to Deborah Shapiro, Alise Reicin, and Alan Nies re: VIGOR. March 9, 2000; www.vioxxdocuments.com/Documents/Krumholz_Vioxx/Scolnick2000.pdf.

167 *The New England Journal of Medicine*: Curfman, G. D.; Morrissey, S.; Drazen, J. M. Expression of concern reaffirmed. *NEJM* 354 no. 11 (March 16, 2006): 1193.

168 a U.S. company: Gottlieb, S. Firm tried to block report on failure of AIDS vaccine. *BMJ* 321 (2000): 1173.

168 The drug company: Nathan, D., and Weatherall, D. Academia and industry: lessons from the unfortunate events in Toronto. *Lancet* 353, no. 9155 (March 6, 1996): 771–72.

170 These ads have been: Gilbody, S. et al. Benefits and harms of direct to consumer advertising: a systematic review. *Qual Saf Health Care* 14 (2005): 246–50, qshc.bmj.com/cgi/content/full/14/4/246.

10. WHY CLEVER PEOPLE BELIEVE STUPID THINGS

174 a classic experiment: Gilovich, T.; Vallone, R.; Tversky, A. The hot hand in basketball: on the misperception of random sequences. *Cog Psych* 17 (1985): 295–314.

176 ingeniously pared-down experiment: Schaffner, P. E. Specious learning about reward and punishment. *J Pers Soc Psych* 48, no. 6 (June 1985): 1377–86.

178 In one experiment: Snyder, M., and Cantor, N. Testing hypotheses about other people: the use of historical knowledge, *J Exp Soc Psych* 15 (1979): 330–42.

179 The classic demonstration: Lord, C. G.; Ross, L.; Lepper, M. R. Biased assimilation and attitude polarization: the effects of prior theories on subsequently considered evidence. *J Pers Soc Psych* 37 (1979): 2098–109.

181 In one, subjects: Tversky, A., and Kahneman, D. Availability: a heuristic for judging frequency and probability. *Cog Psych* 5 (1973): 207–32.

182 Asch's experiments: Asch, S. E. Opinions and social pressure. *Sci Am* 193 (1955): 31–35.

184 the behavior of sporting teams: Frank, M. G., and Gilovich, T. The dark side of self- and social-perception: black uniforms and aggression in professional sports. *J Pers Soc Psych* 54, no. 1 (January 1988): 74–85.

185 It's not safe: The experiments in this chapter, and many more, can be found in *Irrationality* by Stuart Sutherland and *How We Know What Isn't So* by Thomas Gilovich.

11. BAD STATS

186 Let's say the risk: Gigerenzer, G., *Reckoning with Risk* (New York: Penguin, 2003), p. 257.

187 Natural frequencies: Butterworth, B. et al. Statistics: what seems natural? *Science* (May 4, 2001): 853.

187 The other methods: Hoffrage, U.; Lindsey, S.; Hertwig, R.; Gigerenzer, G. Communicating statistical information. *Science* 290, no. 5500 (December 22, 2000): 2261–62.

188 there are studies: Hoffrage, U., and Gigerenzer, G. Using natural frequencies to improve diagnostic inferences. *Acad Med* 73, (1998): 538–40.

200 the same test: Gigerenzer, G. *Adaptive Thinking: Rationality in the Real World.* Oxford University Press: 2000, p. 219.

201 Let's think about: Szmukler, G. Risk assessment: "numbers" and "values." *Psych Bull* 27 (2003): 205–207.

204 a small collection: www.qurl.com/lucia.

12. THE MEDIA'S MMR HOAX

210 In 1957, a baby: Brynner, R., and Stephens, T. D., *Dark Remedy: The Impact of Thalidomide and Its Revival as a Vital Medicine* (New York: Perseus Books, 2001).

211 Many years later: Thalidomide hero found guilty of scientific fraud. *New Scientist* (February 27, 1993).

216 "12 children": Wakefield, A. J.; Murch, S. H.; Anthony, A., et al. Ileal-lymphoid-nodular hyperplasia, non-specific colitis, and pervasive developmental disorder in children. *Lancet* 351, no. 9103 (1998): 637–41.

217 one of the few: e.g., Chess, S. Autism in children with congenital rubella. *J Autism Child Schizophr*, no. 1 (January–March 1971): 33–47.

218 a tenacious investigative journalist: briandeer.com/wakefield/vaccine
     patent.htm 299. "including the BBC": No jabs, no school says labour MP,
     news.bbc.co.uk/1/hi/health/7392510.stm.

225 one survey: Schmidt, K.; Ernst, E.; Andrews, D. N. Survey shows that some
     homoeopaths and chiropractors advise against MMR. British Medical Jour-
     nal 325, no. 7364 (September 14, 2002): 597.

226 32 percent: Hargreaves, I.; Lewis, J.; Speers, T. Towards a better map:
     science, the public and the media, Economic and Social Research Council
     (2003), www.esrc.ac.uk/ESRCInfoCentre/Images/Mapdocfinal_tcm6–
     5505.pdf.

227 peak of the media: Boyce, T., Health, Risk and News: The MMR Vaccine
     and the Media. (New York: Peter Lang Publishing Inc, 2007), p. 307.

227 published a paper: Ibid.

228 not a single one: Durant, J., and Lindsey, N. GM foods and the media. Select
     Committee on Science and Technology, Third Report, Appendix 5, www
     .publications.parliament.uk/pa/ld199900/ldselect/ldsctech/38/3810.htm.

233 a systematic review: Smeeth, L.; Cook, C.; Fombonne, E.; Heavey, L.; Ro-
     drigues, L. C.; Smith, P. G., et al. MMR vaccination and pervasive devel-
     opmental disorders: a case-control study. Lancet 364, no. 9438 (2004):
     963–69.

235 This study was big: Madsen, K. M.; Hviid, A.; Vestergaard, M.; Schendel,
     D.; Wohlfahrt, J.; Thorsen, P., et al. A population-based study of measles,
     mumps, and rubella vaccination and autism. N Eng J Med 347, no. 19
     (2002): 1477–82.

237 "Scientists in America": www.telegraph.co.uk/news/main.jhtml?xml=/news/
     2002/06/23/nmmr23.xml.

240 a very similar study: Afzal, M. A.; Ozoemena, L. C.; O'Hare, A., et al.
     Absence of detectable measles virus genome sequence in blood of autistic
     children who have had their MMR vaccination during the routine child-
     hood immunization schedule of UK. J Med Virology 78, no. 5 (2006):
     623–30.

241 Another major paper: D'Souza, Y., et al. No evidence of persisting measles
     virus in peripheral blood mononuclear cells from children with autism
     spectrum disorder. Pediatrics 118 (October 4, 2006): 1664–75.

243 In some parts: www.westminsterpct.nhs.uk/news/mmr0405.htm; Pearce,
     A., et al. Factors associated with uptake of measles, mumps, and rubella vac-
     cine (MMR) and use of single antigen vaccines in a contemporary UK
     cohort: prospective cohort study. BMJ 336, no. 7647 (2008): 754.

244 a systematic review: Chapman, S., et al. Med J Aust. 183, no. 5 (September 5,
     2005): 247–50. Grilli, R., et al. Cochrane Database of Systematic Reviews 4
     (2001): CD000389.

244 A mischievous paper: Phillips, D. P., et al. N Engl J Med 325 (1991):
     1180–83.

245 systematic quantitative surveys: Schwitzer, G. *PLoS Med* 5, no. 5 (2008): e95.

245 Meanwhile, the incidence: HPA. Confirmed measles mumps and rubella cases in 2007: England and Wales. Health Protection Report 2, no. 8 (2008); accessed April 9, 2008, www.hpa.org.uk/hpr/archives/2008/hpr0808.pdf.

246 Congenital rubella syndrome: Fitzpatrick, M. MMR: risk, choice, chance. *Brit Med Bulletin* 69 (2004): 143–53.

249 epidemic in 2005: Gupta, R. K.; Best, J.; MacMahon, E. Mumps and the UK epidemic. *BMJ* 330 (May 14, 2005): 1132–35.

AND ANOTHER THING

256 "The true cost": www.economist.com/research/Economics/alphabetic.cfm ?letter=O.

FURTHER READING AND ACKNOWLEDGMENTS

I have done my absolute best to keep these references to a minimum, as this is supposed to be an entertaining book, not a scholarly text. More useful than references, I would hope, are the many extra materials available on www.badscience.net, including recommended reading, videos, a rolling ticker of interesting news stories, updated references, activities for schoolchildren, a discussion forum, everything I've ever written (except this book, of course), advice on activism, links to science communication guidelines for journalists and academics, and much more. I will always try to add to it as time passes.

There are some books that really stand out as genuinely excellent, and I am going to use my last ink to send you their way. Your time will not be wasted on them.

Testing Treatments by Imogen Evans, Hazel Thornton, and Iain Chalmers is a book on evidence-based medicine specifically written for a lay audience by two academics and a patient. It is also free to download online from www.jameslindlibrary.org. *How to*

Read a Paper by Professor Trisha Greenhalgh is the standard medical textbook on critically appraising academic journal articles. It's readable and short, and it would be a bestseller if it weren't unnecessarily overpriced.

Irrationality by Stuart Sutherland makes a great partner with *How We Know What Isn't So* by Thomas Gilovich, as both cover different aspects of social science and psychology research into irrational behavior, while *Reckoning with Risk* by Gerd Gigerenzer comes at the same problems from a more mathematical perspective.

Meaning, Medicine and the "Placebo Effect" by Daniel Moerman is excellent, and you should not be put off by the fact that it is published under an academic imprint.

There are now endless blogs by like-minded people that have sprung from nowhere over the past few years, to my enormous delight, onto my computer screen. They often cover science news better than the mainstream media, and the feeds of some of the most entertaining fellow travelers are aggregated at the website badscienceblogs.net. I enjoy disagreeing with many of them—viciously—on a great many things.

And last, the most important references of all are to the people by whom I have been taught, nudged, reared, influenced, challenged, supervised, contradicted, supported, and, most important, entertained. They are (missing too many, and in very little order): Emily Wilson, Ian Sample, James Randerson, Alok Jha, Mary Byrne, Mike Burke, Ian Katz, Mitzi Angel, Robert Lacey, Chris Elliott, Rachel Buchanan, Alan Rusbridger, Pat Kavanagh, the inspirational badscience bloggers, everyone who has ever sent me a tip about a story on ben@badscience.net, Iain Chalmers, Lorne Denny, Simon Wessely, Caroline Richmond, John Stein, Jim Hopkins, David Colquhoun, Zoe Pagnamenta, Chantal Clarke, Sarah Ballard, Shalinee Singh, Catherine Collins, Matthew Hotopf, John Moriarty, Alex Lomas, Andy Lewis, Trisha Greenhalgh, Gimpy, shpalman, Holfordwatch, Positive Internet, Jon, Liz Parratt, Patrick Matthews, Ian Brown, Mike Jay, Louise Burton, John King, Cicely Marston, Steve Rolles, Hettie, Mark Pilkington, Ginge Tulloch, Matthew Tait, Cathy Flower, my mum, my dad, Reg, Josh, Raph, Allie, and the fabulous Amanda Palmer.

INDEX